RCM—GATEWAY TO WORLD CLASS MAINTENANCE

RCM—GATEWAY TO WORLD CLASS MAINTENANCE

Anthony M. Smith and Glenn R. Hinchcliffe

ELSEVIER
BUTTERWORTH
HEINEMANN

AMSTERDAM • BOSTON • HEIDELBERG • LONDON • NEW YORK
OXFORD • PARIS • SAN DIEGO • SAN FRANCISCO • SINGAPORE
SYDNEY • TOKYO

Elsevier Butterworth–Heinemann
200 Wheeler Road, Burlington, MA 01803, USA
Linacre House, Jordan Hill, Oxford OX2 8DP, UK

Library of Congress Cataloging-in-Publication Data

Smith, Anthony (Anthony M.)
  RCM : gateway to world class maintenance / Anthony Smith and Glenn R. Hinchcliffe.
  p. cm.

  ISBN-13: 978-0-7506-7461-4   ISBN-10: 0-7506-7461-X

  1. Plant maintenance. 2. Reliability (Engineering) 3. Maintainability (Engineering) I. Hinchcliffe, Glenn R. II. Title.

  TS192.S655 2003
  658.2—dc22   2003062766

ISBN-13: 978-0-7506-7461-4
ISBN-10: 0-7506-7461-X

British Library Cataloguing-in-Publication Data
A catalogue record for this book is available from the British Library.

For information on all Butterworth–Heinemann publications visit our website at www.bh.com

Transferred to Digital Printing 2010

To our wives, Mary Lou and Susan, whose support and belief in our journey made it all possible.

CONTENTS

FOREWORD

Anthony M. (Mac) Smith and I share a common source for our beliefs in the merits and benefits of Reliability-Centered Maintenance (RCM). Like me, Mac was the beneficiary of generous mentoring by Thomas D. Matteson, Vice President for Maintenance Planning at United Airlines (UA) in the 1970s. Tom was an early contributor to and champion of the methodology we now call RCM. In 1971, Tom arranged for UA personnel (F. Stanley Nowlan, Howard F. Heap and others) to meet with representatives of the U.S. Navy office where I worked. The meeting agenda was to review the work and results of efforts of the 747 Maintenance Steering Group and the newer Air Transport Association R & M Subcommittee. We had already studied the documents that these groups, which included members from UA, had produced (MSG-1—Handbook: Maintenance Evaluation and Program Development, and MSG-2—Airline/Manufacuturer Maintenance Program Planning Document). This was seven years before the seminal report for the U.S. Department of Defense by Nowlan and Heap was published as a book titled Reliability-Centered Maintenance.

What the Navy representatives needed was some indication that what was proposed in earlier documents could be effectively implemented to benefit the USA's already very reliable fleet of nuclear-powered ballistic missile submarines. These ships were indisputably the least vulnerable of the triad of systems (including land-based strategic missiles and bomber aircraft) that formed U.S. deterrent forces. The problem was that maintenance of these ships was becoming prohibitively expensive in light of other priorities for defense dollars which, at the time, included fighting the war in Vietnam. It was assumed (correctly) that defense funding, even for high-priority strategic systems, would become even harder to acquire for years after that war ended. We needed a financial miracle!

If ever there was a profound moment of revelation in maintenance and its relation to reliability, the 1971 meeting with UA personnel was it for us. From then on, the principles and methods discussed in that meeting became a centerpiece of the plan we developed and implemented for changing the way the Navy had maintained ships throughout its entire history. As a result, the 31 ships in the initial group under the revised maintenance program cost significantly less to maintain (estimated at $12.7 billion by 1988, ~15% of the previously projected total life-cycle maintenance cost). In addition, they were available for their deterrent missions about 5% more than originally planned. Some of these ships were also able to serve reliably as long as 8 to 10 years beyond their expected lives of 25 years. RCM became an important influence on the succeeding (Trident) class of nuclear-powered ballistic missile submarines that will continue to operate well into the 21st century.

In 1981, the organization where I worked, then called the Submarine Monitoring, Maintenance and Support Office (SMMSO), began applying RCM to nuclear-powered attack submarines. One of the beneficiaries, for example, was the real USS Dallas (SSN 700). This was the U.S. attack sub made famous in Tom Clancy's novel and the motion picture entitled The Hunt for Red October. The results from an operations and maintenance standpoint were, and continue to be, much the same for all ship classes subjected to RCM as it was for missile subs—lower life cycle total costs and sustained or improved reliability and availability.

It was also in 1981 that Mac Smith and Tom Matteson approached the Electric Power Research Institute (EPRI) key personnel with a recommendation to apply RCM methodology to nuclear utility plants. This was just two years after the reactor core meltdown accident at one of Pennsylvania's Three Mile Island nuclear-powered electricity generating plants (TMI-II). EPRI had been under heavy pressure from its subscriber utilities to focus on more "practical" projects that could yield immediate benefits, rather than on research and development that might not produce useable results for decades. RCM offered many near-term benefits, including those affecting economics and reliability. It was also felt that application of the methodology could help rebuild public confidence in commercial nuclear power and counter criticism of its very vocal opponents.

Results of RCM pilot projects led by Mac Smith and Tom Matteson had much to do with the focus by utility personnel on the methodology. These pilot studies were done at Florida Power & Light (FPL) Turkey Point Nuclear Generating Station and Duke Power Company McGuire Nuclear Station. EPRI and the utilities co-sponsored these projects as a result of a series of convincing presentations by Mac and Tom starting with the one, mentioned above, in 1981. The extensive reports of results of RCM analyses, published in 1985 and 1986, created a sensation in the utility industry and supporting companies. Almost overnight, many consulting firms began offering RCM-related and "RCM-like" services and products. RCM became the "in" thing to do in the electric utility world. EPRI formed an RCM Users' Group for information exchange on the subject. Many electric utility executives took notice and began to prod their subordinates for action on the issues related to RCM.

As time passed, an impact of RCM that became important to nuclear utilities was how federal government regulators (Nuclear Regulatory Commission (NRC) personnel) viewed its application. In addition, a utility industry "watchdog" organization (Institute for Nuclear Plant Operations—INPO) was created in response to the accident at TMI-II. An unwritten goal of INPO was to help "save" the commercial nuclear power industry from demise due to its apparent lack of discipline in operations and maintenance (O & M). INPO began to recognize RCM as one of many means to help nuclear utilities save themselves.

One of the key questions the regulators began to ask and INPO industry advisors tried to help utilities answer was "What is the basis for the maintenance programs at the nuclear plants your organization operates?" If utility personnel had a comprehensive set of RCM analyses for critical systems and had adequately implemented the results, they usually could give an acceptable answer. If not, it was more difficult to convince NRC inspectors and INPO advisors that commercial nuclear plant operators and maintainers had their act together.

Many NRC personnel and an even greater percentage of INPO personnel (including top executives of both organizations for many years) were former nuclear submariners. They all had hands-on O & M experience and understood that consistency of performance resulted, in part, from a solidly based maintenance program such as they had been responsible for on submarines. By the mid-1990s their influence resulted in the so-called "Maintenance Rule" becoming a matter of law with which nuclear utilities had to comply. Incorporated into the Code of Federal Regulations was the requirement for commercial nuclear power plants to have a "...well defined and effective program to assure that maintenance activities are conducted to preserve or restore the availability, performance and reliability of plant structures, systems and components." Tom Matteson also played a part here. In 1990, I found myself working beside him as a "peer reviewer" for NRC. We were helping commission engineers and their principal contractors to define how RCM (called Risk Focused Maintenance—RFM by NRC) could be used in conjunction with nuclear plant Probabilistic Risk Assessments to meet the requirement established by the regulation mentioned above.

When Glenn Hinchcliffe (and others) supporting fossil (oil and gas) powered electricity generating plants within FPL became aware of the results obtained in the pilot project at Turkey Point Station, they began to wonder whether or not their 27 fossil power stations could benefit from application of RCM. After studying the results and the basic methodology, Glenn became a believer in the value of applying the discipline and logic of RCM to systems of all types. He was appointed FPL's leader responsible for a team of 12 experts in boilers, turbines, generators, motors, pumps, and other major components and systems to apply RCM to all their fossil plants. He arranged to have Mac Smith hired to consult. Guided by Mac and Glenn, FPL personnel developed their own training programs and RCM presentation formats based on the "classical" RCM approach articulated in the book by Nowlan and Heap.

In 1989, the work led by Glenn Hinchcliffe in the field of RCM was a key factor in FPL's winning the coveted Deming Application Prize (for Total Quality Control) awarded by the Japanese Union of Scientists and Engineers. The prize was named for Dr W. Edwards Deming, the much revered American who is credited with being the key person responsible for lifting Japan's manufacturing expertise and competitiveness to world class status. FPL was the first company outside of Japan to win this prestigious award. Only a few others outside of Japan have won it since.

From the late 1980s to 1992, Mac Smith was developing his own milestone book entitled *Reliability-Centered Maintenance*. McGraw-Hill published it in 1993. As he had acquired considerable RCM practitioner experience at FPL, Glenn Hinchcliffe contributed significantly to its content and provided review comments to help Mac with editing. By then the original report by that title, written by Nowlan and Heap, was out of print. Although all of its central principles were (and still are) valid, its focus on commercial aircraft application made it difficult for maintenance personnel in other, less glamorous, industries to "sell" management on adopting them. Mac's book (and others like it) was badly needed to spread the good news to stakeholders of all organizations that could benefit from competently executed RCM. Now, for this latest effort, Glenn shares the full responsibility of co-authorship to assure that its content provides valid, meaningful guidance to those that invest time reading and studying it. I believe he (and Mac) succeeded admirably in making it well worth the time to do this.

The co-authors shared drafts of chapters of this book with anyone interested in reading them. Mac and Glenn solicited feedback and recommendations for improvement from many sources. The result for the reader is refinement and sharper focus of the co-authors' opinions and experience. Of special note, they requested clearance from former clients to publish the case studies provided in Chapter 12. The studies provide a good cross-section from government, utility, and manufacturing sectors. These are not anonymous examples; the clients are named. Each reflects real-world findings and results that the client certified were accurate before their publication.

In particular, readers' attention should focus on the statistical results provided in the "PM Task Similarity Profiles (by Failure Mode)" in the case studies. These reflect what should happen for critical, problematic systems subjected to competent RCM analysis. If whatever you did previously to upgrade your maintenance effort hasn't yielded results like these, you haven't achieved what's possible—or prudent.

A good self-administered test for maintenance and reliability practitioners includes the following questions. "Can I thoroughly document the basis for my organization's maintenance and reliability program for systems that affect throughput, quality, cost, and environmental safety?" "Could I defend that basis before a serious management inquiry, in front of regulators that could affect my plant, or in a court of civil or criminal law?" If you can't answer these questions adequately, you should investigate how RCM, practiced as outlined in this book, can help.

In October 2002, at the Society for Maintenance and Reliability Professionals 10th Annual Conference, I had the pleasure of seeing, listening to and socializing with Mac Smith, again, face-to-face. We have exchanged phone calls and e-mails often in recent years, but this was special. Mac was co-author of two of the papers presented (one with Glenn and others) and led a full day's RCM workshop (again with Glenn). No other presenters were as visible during that conference as they were.

During a break between conference sessions, I saw Mac in intense conversation with the other giant (currently) in the field of RCM, John Moubray (mentioned in Chapter 4 of this book). They were meeting for the first time. I desperately wanted to listen in on their private discussion, but didn't dare intrude. I thought to myself that someone should take a photo of the two, together, during that conference. Their joint presence has historical significance for our profession. They started their journeys into RCM about as far away from each other as two individuals could on this planet—John from South Africa (but now a resident of the USA) and Mac from northern California. However, each traces the roots of his knowledge about the subject to RCM pioneers who worked together at United Airlines. John credits the late Stan Nowlan; Mac does the same for Tom Matteson. They share the common goal of ensuring RCM is executed and defined properly for the benefit of all. Smith and Moubray are now, in my opinion, the two professionals who have had the most influence on improving the practice of maintenance and reliability, worldwide, in the past decade. My hopes, prayers, and expectations are that they will continue to do so well into this new millennium. There's still much to do.

So, for maintenance and reliability practitioners, supporting managers and executives in their organizations, this book has a lot to offer. All should take the time to study its contents and apply lessons learned that Mac Smith and Glenn Hinchcliffe have acquired—many times the hard way, through failure (see Chapter 9), as well as success.

Jack R. Nicholas, Jr, P.E., CMRP
Submarine Monitoring, Maintenance and Support Office (1971–88)
Currently—CEO Maintenance Quality Systems LLC
Gettysburg, PA
December 2002

PREFACE

Ten years ago, my book "Reliability-Centered Maintenance" was published (Ref. 1). Over the course of this decade, my horizons in coping with practical maintenance issues have been significantly expanded. Also, for the past 15 years, I have had the privilege of working with Glenn Hinchcliffe—first as a client and later as an Associate. Glenn has been a valuable assistant to me on my learning journey, and I am pleased to have his valuable insights as a co-author for this book.

In Ref. 1, I basically supported the view that RCM offered what is perhaps the best way to develop a maintenance improvement program. Now, a decade later, Glenn and I will unequivocally state that RCM is most definitely the best way to do this. And, to go one step further, we have also formulated our view of the five key steps required to achieve a World Class Maintenance Program. RCM plays the pivotal role in this scenario (see Sec. 1.5).

Our learning experiences come almost exclusively from the opportunities that we have had in knowing and working with people like you who are yourselves believers in what RCM can achieve, or are just starting your journey to maintenance excellence. These associations and the learning that resulted have been the experiences of a lifetime for us. And now we want to share some of this accumulated experience with you.

This book contains much of the core RCM process that was described originally in Ref. 1. But the discussions have been augmented in several places with additional explanations and clarifications where we have found problems with people not finding it easy to grasp certain features in the maintenance and RCM process.

The RCM methodology that was originally learned from Tom Matteson has not changed. It was right then, and it is still right now. But we have learned a great deal about how to do it better, and how to avoid the more significant pitfalls. Much of this new book is devoted to such information.

We are especially pleased with the assistance that was provided to us by seven of our clients in order to present seven success stories (Case Studies) to illustrate what RCM can do for you when you do it the right way (see Chapter 12). Totally new chapters are also devoted to topics such as Implementing RCM (Chapter 8), Lessons Learned (Chapter 9), The Living RCM Program (Chapter 10), and RCM Analysis Supporting Software (Chapter 11). Of course, the basic methodology for performing Classical RCM is here (Chapter 5) with a simple illustration of its application to a swimming pool (Chapter 6). There is also a new chapter devoted to Alternative Analysis Methods that may be useful in certain circumstances with the more well-behaved systems, including an Abbreviated Classical RCM™ process (Chapter 7). There is also an Appendix devoted to a model for analyzing the economics of preventive maintenance that is contributed by Dr David Worledge.

I hope that the book proves useful to you in your professional work. Let me know what you think.

Anthony M. (Mac) Smith
San Jose, California
December 2002
E-mail: amsassoc@sbcglobal.net

ACKNOWLEDGEMENTS

Your authors have been on an unbelievably exciting journey for the past 22 years. Our professional careers have provided us with the opportunity to influence and reshape much of the thinking that drives the maintenance philosophy across the U.S. industry. During this journey, we have had the honor to associate with literally hundreds of maintenance personnel, at all levels in the organization. These people have shared with us their experiences and own personal feelings about what is right (and wrong) with industry O & M practices. These associates have provided us with much of the background and practical material that allowed us to write this book. For all of this, we are profoundly grateful to each and every one of you.

A major milestone occurred since the publication of Ref. 1 that has had a profound and beneficial impact on our work—namely, the development and application of the "RCM WorkSaver" software. Of course, the real beneficiaries of this new software have been our clients, and, without exception, they have been very pleased to have this capability added to their RCM programs. Our thanks to JMS Software and Nick Jize, Jim McGinnis and Joe Saba who created "RCM WorkSaver" to our specifications.

Producing a manuscript like this takes a great deal of skill and patience to do all of the word processing of the draft material. In all of this, we have had the consistent support of Ann Mullen and Paul Bernhardt who have been instrumental in assuring that we got from point A to point B in producing this manuscript. Many thanks to both of you.

Finally, we are indebted to the staff at Elsevier for their efforts in accepting and publishing this book in such a professional manner.

Please Note: World Class Maintenance® is a registered trademark of HSB Reliability Technologies and is used throughout this text with their permission.

DISCLAIMER

Information contained in this work has been obtained by Elsevier from sources believed to be reliable. However, neither Elsevier nor its authors guarantee the accuracy or completeness of any information published herein and neither Elsevier nor its authors shall be responsible for any errors, omissions, or damages arising out of use of this information. This work is published with the understanding that Elsevier and its authors are supplying information but are not attempting to render engineering or other professional services. If such services are required, the assistance of an appropriate professional should be sought.

1

WORLD CLASS MAINTENANCE (WCM)—OPPORTUNITY AND CHALLENGE

Until the late 1970s timeframe, product development and manufacturing engineering were the dominant technical disciplines in the U.S. industrial community, with operations and maintenance (O&M) occupying a back seat in the priority of corporate success strategies. The past two decades, however, have seen this picture shift rather dramatically to a situation where O&M is now a peer with the development and manufacturing disciplines. There are compelling reasons for this, not the least of which is the decisive role that O&M now plays in issues ranging from safety, liability, and environmental factors to bottom-line profitability. With O&M now center stage, Preventive Maintenance (PM) optimization (i.e., World Class Maintenance—WCM) is providing never-before-seen opportunities and challenges to the O&M specialists.

Some of these challenges come in the form of various maintenance problems that currently have a great deal of commonality across our industrial system. Twelve such problems are briefly discussed to indicate some specific dimensions of the challenge before us. Many people share the view that the Reliability-Centered Maintenance (RCM) methodology offers the best available decision strategy for PM optimization and WCM. The authors strongly share this view. And that is what this book is all about.

1.1. SOME HISTORICAL ASPECTS

The title of this introductory chapter contains two keys that set the stage for the continuing theme throughout this book—challenge and opportunity. Let's step

1

back for a moment from the everyday pressures and excitement that are typically associated with a plant or facility operation, and look at what these words might mean to us.

Since the end of World War II, the growth of the U.S. industrial complex has been dominated by two factors: (1) technical innovations, which have led to a plethora of products that were mere dreams in the pre-World War II period; and (2) volume production, which has enabled us to reach millions of customers at prices well within the reach of virtually every consumer in the United States and its international peers. From a motivational point of view, product development and design, as well as the manufacturing engineering that provides for mass production capabilities, have been the "darlings" of the engineering world during the 1960s, '70s and '80s. And, as a result, the bad news is that O&M was often relegated to a "necessary evil" role with all of the attendant problems of second-class citizen status when it came to research and development (R&D) projects, budget requests, manpower allocations, and management awareness. Somehow, the reasoning goes, those good people in the trenches seem to keep things running—so they get the token pat-on-the-back and another year's supply of bailing wire and chewing gum to keep it all together. A bit of an exaggeration? Maybe, but some of you are probably relating to this scenario in some fashion.

Well, here is the good news. Things have been rapidly changing over the past two decades, as you have most likely noticed. The reasons for this are well recognized—environmental concerns, safety issues, warranty and liability factors, regulatory matters, and the like. But most of all, as plant and equipment age and global competition becomes a way of life, management has realized that O&M costs are (or could be) "eating their bottom-line lunch." There has been a very positive shift in management concern and awareness about both the cost and technical innovation of O&M policies, practices, and procedures. In fact, we might even go so far to say that O&M now rightfully occupies a peer position with product development and manufacturing engineering in many product areas, and the need for World Class Maintenance programs is receiving top-level corporate attention.

And that is where the challenge and opportunity come into this picture. Make no mistake, O&M is now in the center-stage spotlight. What are we going to do about it?

It is hoped that this book will help you to answer that question with regard to one major aspect of the O&M challenge—how to get the most from the resources committed to the plant or facility preventive maintenance program. It is suggested that a viable WCM strategy can be PM optimization via the use of the Reliability-Centered Maintenance methodology. In subsequent chapters, our aim is to describe, in some detail, just exactly how this can be achieved.

1.2 SOME COMMON MAINTENANCE PROBLEMS*

With the O&M spotlight coming center stage, it is instructive to look at some industry-wide maintenance history of the past three decades, especially with respect to some of the more classic maintenance problems that we need to address. It is recognized that the list of problems discussed here is not all-inclusive, nor are these problems necessarily common to everyone. But they are thought to represent a mainstream of experiences that have been observed with sufficient frequency to warrant their attention here.

1. *Insufficient proactive maintenance.* This problem clearly heads the list simply because the largest expenditure of maintenance resources in plants typically occurs in the area of corrective maintenance. Stated differently, the vast majority of plant maintenance personnel operate in a reactive mode, and in some instances plant management actually has a deliberate philosophy to operate in such a fashion. It is interesting to note that, in the latter case, the end product from such a plant usually has the highest unit cost within the group of peers producing the same product. A major contributor to the unit cost thus accrues from a combination of the high cost to restore plant equipment to an operable condition coupled with the penalty associated with lost production. Since simple arithmetic readily demonstrates this fact, it is really quite surprising that this reactive environment continues to occur. Yet, to one degree or another, it seems to be a commonplace situation.

2. *Frequent problem repetition.* This problem, or course, ties in rather directly with the preceding. When the plant modus operandi is reactive, there is only time to restore operability. But there is never enough time to know why the equipment failed, let alone enough time or information to know how to correct the deficiency permanently. The result is that the same problem keeps coming up—over and over. This repetitive failure problem is often discussed in terms of root cause analysis, or more appropriately the lack thereof. Unless we understand why the equipment failed and act to remove the root cause, restoration to service may be a temporary measure at best, and the cycle not only continues but is reinforced.

3. *Erroneous maintenance work.* Humans make mistakes, and errors will occur in maintenance activities (both preventive and corrective). But what is a tolerable level of human error in a maintenance program? Is it 1 error in 100 tasks, 1 in 1000, or 1 in 10? The answer could depend on the consequence realized from the error. If you are a frequent flyer, you would like to believe that both maintenance errors and pilot errors

---

*Inputs provided by the Electric Power Research Institute (EPRI) and the EPRI RCM Users Group are gratefully acknowledged as partial source material for this discussion.

are less than 1 in 1 million! (In terms of catastrophic errors, they are in this range.) But let's think about this in economic terms—the cost of another corrective action and attendant loss of plant production. Most plant managers wish to have that 1 in 1 million statistic, but seem to believe that reality is more like 1 in 100. There is strong evidence, however, that human error which occurs during intrusive-type maintenance actions is the cause of about 50 percent of plant forced outages, and that some form of human error might be occurring in some locations in one of every two maintenance tasks that are performed. Surprised? Check your own records; they may be a real eye-opener. Or check your automobile service records—how many times did you return to correct something that did not exist on the initial shop visit? More is said about the concern over erroneous maintenance work in Secs. 2.5 and 4.2.

4. *Sound maintenance practices not institutionalized.* One way to solve the human error problem is, for starters, to know the practices and procedures that can assure that mistakes are not made—and then to institutionalize them in the everyday work habits at the plant. Collectively, industry has a great deal of knowledge and experience on how equipment should be handled (often called Best Practices)—how it should be removed from the plant, torn down, overhauled, reassembled, and reinstalled. Individual plants are usually informed on only a small percentage of this collective picture. Worse yet, what is known is all too infrequently committed to a formalized process (procedures, training, etc.) and transferred across corporate barriers to other plants in the organization. Since virtually all of the authors' experience deals with Fortune 100 companies, we can attest to the veracity of this problem.

5. *Unnecessary and conservative PM.* At first glance, one might feel that this problem is in conflict with item 1 above. While the need for more PM coverage is clearly an appropriate issue, there is a parallel need to look at the PM that is currently performed in terms of "Is it right?". Historical evidence strongly suggests that some of our current PM activities are, in fact, not right. In some cases, PM tasks are totally unnecessary because they have little, if any, relationship to keeping the plant operational. (In later chapters, we will see that this problem relates to the lack of what we call task "applicability.") It is not uncommon to examine a plant PM program, and find 5 to 20 percent of the existing tasks could be discarded and the plant would never know the difference. The trouble is that most plants never revisit PM tasks with this question in mind. That is, they have no PM "Living Program" (see Chapter 10 for a comprehensive discussion on this topic). A second form of this problem is where the PM task is right, but too conservative. This problem is usually associated with task frequency (i.e., the frequency requires the PM action too often). This seems to be

especially true of major overhaul tasks where there is some substantial evidence to suggest that 50 percent or more of the PM overhaul actions are performed prematurely. See Sec. 5.9 for additional insights on this issue.

6. *Sketchy rationale for PM actions.* Did you ever ask the maintenance manager why some PM task is being performed? Did you receive a credible response? Could the response be supported with any documentation that could reasonably reconstruct the origins of the task (other than "the vendor told us to do it"—see item 8 following)? Unfortunately, the absence of information on PM task origin or any documentation to clearly trace the basis for plant PM tasks is the rule and not the exception. Perhaps one might suggest that this is not always unacceptable. If maintenance costs (PM + CM) were low and still decreasing, and if plant forced outages were virtually nonexistent, one might allow that this could be the case. But neither of these factors is sufficient for us to ignore the issue of why we do a PM task, nor to forego the ability to record the basis for such actions in appropriate documentation. For example, the Federal Aviation Administration requires an approved and documented basis for PM as a requirement for aircraft Type Certification (i.e., approval to build and sell the airplanes), a requirement that has been in place for several decades. More recently, nuclear power plants have also adopted a more formalized PM process as one element of their implementation of the Maintenance Rule that was issued by the Nuclear Regulatory Commission.

7. *Maintenance program lacks traceability or visibility.* If the plant does not perform routine analyses of equipment failure cause, and is remiss at recording the basis of PM actions, then at least two significant areas have been defined where visibility and traceability of decisions and actions are missing. But the problem goes beyond this in many situations, and we refer here to the lack of definitive information in the Computerized Maintenance Management System (CMMS). Frequently, there is no traceable record of PM actions and costs to be found anywhere except in the heads or desk drawers of the plant staff. If they leave, the plant memory walks out the door with them. In today's world of efficient and inexpensive computer systems, complete with customized software, there seems to be little excuse not to have good plant records on what was done (or is scheduled to be done), and why management decided on certain definitive strategic and tactical courses of action.

8. *Blind acceptance of OEM inputs.* The original equipment manufacturer (OEM) almost always provides some form of operations and maintenance manual with the delivered equipment. From a PM point of view, two problems develop with this input. First, the OEM has not necessarily thought through the question of PM for the equipment in

a comprehensive and cost-effective fashion. Often, the OEM PM recommendations are last-minute thoughts that tend to be aimed at protecting the manufacturer in the area of equipment warranty (this is the origin of many conservative PM tasks). Second, the OEM sells equipment to several customers, and these customers operate that equipment in a variety of different applications—for example, cyclic rather than steady state, very humid rather than dry ambient conditions, etc. The OEM usually designs the equipment with some operational variability in mind, but rarely does the vendor ever specifically tailor the equipment to your special needs. Yet the basis for many PM programs is the blind acceptance of OEM PM recommendations as the best course of action to pursue—even though the OEM recommendations are conservative and not necessarily applicable to the plant's operating profile.

9. *PM variability between like or similar units.* Within a given company, it is likely that multiple plant locations are involved in production and, in some instances, each plant may even have multiple units at each location. The utility industry typifies the latter situation where two or more power generation units are frequently located at each plant site. These multiple plant or unit situations are likewise composed of production facilities that are often virtually identical from site to site or, at the very least, contain a wide spectrum of equipment that is identical or highly similar. Under these circumstances, it would seem reasonable to assume that their PM programs share this commonality in order to standardize procedures, training, spare inventories, etc. to capitalize on the obvious cost savings that can be achieved. Unfortunately, this is not a good assumption; more often than not we find that each plant location tends to be its own separate entity with many of its O&M characteristics different from those of its sister plants within the company. It is not clear why corporate management allows this to occur, but at the plant level of organization there appears to be a strong feeling of "I'm not like them" and "We know what's best for us" attitudes driving this lack of commonality. Across a given industry composed of multiple companies with similar product lines, this situation becomes even more pronounced (and perhaps a bit more understandable due to varying corporate cultures and competitive restrictions on information exchange).

10. *Ineffective use of predictive maintenance technology.* This new area of maintenance technology has been evolving for several years and is usually described under the name of predictive maintenance (PdM). It is also described with titles such as condition monitoring, condition-based maintenance, monitoring and diagnostics, and performance monitoring. All of these names are intended to describe a process whereby some parameter is measured in a nonintrusive manner and either trended over time or alarmed at some predetermined limit, the

said parameter being one with a direct relationship to equipment health, or at least to some specific aspect of equipment health. Clearly, this process has the potential for significant payoff when it can be used to tell us when it is necessary to perform some maintenance task, thus precluding both unnecessary as well as premature intrusive preventive maintenance actions that otherwise would occur. Some of this technology is fairly sophisticated (e.g., vibration analysis on rotating machinery) and some of it is fairly simple (e.g., pressure drop across a filter). To a large extent, much of this technology is being introduced into our plants and facilities. But where a plant has a predictive maintenance program, more often than not its focus is on the deployment of the sophisticated, not the simple technologies. Also, this deployment is fairly global and not always directed at critical functions where the return on investment (ROI) would be significant.

11. *Failure to employ the 80/20 rule.* Our experience across several areas of U.S. industry has indicated that the majority of O&M managers and their staff do know about the 80/20 rule, and have a reasonable grasp of its meaning. This rule states that 80% of an observed effect tends to reside in 20% of the available source. For example, 80% of carpet wear is found in 20% of the available carpet area—because this is where the traffic occurs. So, in a plant, 80% of the reactive maintenance and lost production costs are likely to be located in 20% of the plant's systems—the so-called bad-actor systems. So it would seem rather logical to find that these managers would utilize this rule to allocate resources and to focus their priorities on the bad-actor 20%. Surprisingly (to us), we rarely find this to be the case, and must conclude that there are lost opportunities for cost-effective decisions and actions occurring daily in our plants and facilities.

12. *Absence of long-range commitments.* Any student of management practices would probably take immediate exception to the foregoing statement because strategic (long-range) planning is pretty well ingrained in our industrial culture. But note that the difference of importance here lies in the words planning vs. commitment. We find this to be especially true in the O&M world where top and middle level management is focused on the quarterly results, more commonly known as the Wall Street syndrome. We hear phrases like "low hanging fruit" and "short-term payback" which all too frequently describe the mindset for approving the commitment of resources to achieve O&M improvements. The problem here, of course, is that the ability to achieve noticeable improvements is rarely "short-term" because all of the "low hanging fruit" was plucked long ago. We suggest that a fundamental shift in mindset toward long-range commitments needs to occur.

The opportunity and challenge mentioned in the title of this chapter fall to a large degree in doing something constructive to resolve the above twelve issues.

1.3 PROLIFERATION OF "SOLUTIONS"

1.3.1 The Acronym Parade

As a general observation, the authors have found an increasing sensitivity in the past decade throughout industry and government to the need to improve O&M practices. But conversely, we rarely find the O&M decision-makers directly addressing one or more of the issues listed above in Sec. 1.2. Rather, it is quite common to hear management, in one form or another, make statements like "we know that improvement is needed, but where and how do we start?" or "downtime is killing us, but we can't seem to stop the bleeding." In response to these general concerns, a proliferation of "solutions" has emerged in recent years, which portend to offer various programs and methodologies that will produce quantum improvements. Many of them are couched in catchy acronyms. For example, how many of these do you recognize, and maybe have even tried?:

CBM	RAV	TPM
ELM	RCFA	TPR
EVA	SMW	TQM
JIT	TPE	WIIFM
OEE		

In addition to this alphabet soup of possibilities, perhaps your organization has also been exhorted to re-engineer the whole nine yards. This is a radical revolutionary suggestion which almost always results in organizational chaos and massive human disruption. We find that evolution, not revolution, in the pursuit of change is a much preferred business approach. Other popular suggestions involve the Team Concept, Employee Empowerment, Right Sizing (a frequent euphemism for reduction in force), Benchmarking, and above all, a program for Continuous Improvement. It is hard to argue with the last item, but the burning question is "How do you continuously do it?" (see Chapter 10 for some ideas).

The authors have had some very direct and personal experience with several of these solutions that have been tried by our clients. One particular experience involved our personal participation in an RCM project during the attempted "solution" implementation of Total Quality Management (TQM) at Florida Power & Light. If you are not familiar with this story, you will find Ref. 2 to be fascinating reading. While it is true that FP&L earned the American equivalent of the Japanese "Deming Award" for its effort, it is also true that they incurred the wrath of their employees, customers, shareholders, and the Florida PUC for doing it, and, coupled with other acquisition miscues (primarily Colonial Penn Insurance Company), experienced severe financial problems in the 1980s! Incidentally, the author of Ref. 2 is the ex-CEO of FP&L who was essentially "retired" for his role in pursuing and achieving the award.

Another interesting and somewhat popular solution is Total Productive Maintenance (TPM). In Ref. 3, a leading expert in the TPM methodology provides some sage insights into this solution. Hartman says "At least every second attempted installation of TPM results in failure." His paper then outlines an agenda for a TPM feasibility study which contains 32 bullets of "to-do" actions, followed by a 12-point program which culminates in an award that testifies that your plant is world class. You might get mentally exhausted just reading Ref. 3, but might also question just why these twelve points are the correct ingredients for world class status.

And, speaking of TPM, the measurement of OEE is touted as one of its major cornerstones, which signifies that you have arrived when OEE is 85 percent or better (Refs. 3 and 5). The problem comes when you try to figure out how to establish a data system that will collect, analyze, and then produce credible OEE measurements that are actively employed in management decision-making. It has probably occurred but we haven't witnessed a truly successful working TPM program in our 20-plus years in the maintenance field. And it seems very difficult to really justify the 85 percent figure—why not 75 percent or 95 percent?

At least one maintenance practitioner believes that increased complexity in the conduct of your maintenance business is a prerequisite in order to achieve a World Class Maintenance status (Ref. 4). The list of "solutions" above would certainly tend to support that belief—if they consistently worked. At the risk of being lone voices in the wilderness, we believe that just the opposite is true; that the key to World Class Maintenance status is couched in a rather simple solution. We discuss our thoughts in that regard in Secs. 1.4 and 1.5 below. We cannot emphasize too strongly, however, the importance that we attach to the notion of SIMPLE. All too often, O&M organizations are heading down the path of very complex organizational experiments, overnight attempts at cultural change, and unrealistic expectations of dramatic and highly visible payoffs for a relatively small and short-term investment. This has been a formula for less than successful outcomes. In the trenches, the troops (i.e., those folks who really determine what does and doesn't work) call these programs "the flavor of the day." We believe that our experiences, expressed in the following sections, suggest a rational logic and workable formula to achieve meaningful O&M optimization.

1.3.2 Benchmarking and Best Practices—Help or Hindrance?

A favorite corporate exercise these days is to define where you are vis-à-vis others in your related industry, and especially to know in some detail how you stack up to your competition. This exercise is usually referred to as Benchmarking. The basic thrust is to develop credible benchmarks (i.e., measurable parameters) that describe how your company is doing, measure them, and then compare your company's benchmarks against others to see if you are above or below par. There are, however, some fundamental problems with this seemingly simple concept. For example:

1. What specific parameters are indicative of your company, product, process, and/or customer base that truly define your position?

2. When you benchmark your position against others, do you obtain meaningful results if those results do not characterize a direct competitor? For example, if you build airplanes, do you learn useful information if you benchmark against an automotive manufacturer? And if Company A wants to benchmark against Company B, how do you ascertain that Company B is good enough to teach you some positive lessons. In fact, if you are not careful, you might just try to emulate something that is, in reality, quite damaging within your culture!

3. Clearly, your primary intent is to benchmark against a competitor. But do you really have access to data about your competitor? If you do, is it credible and sufficiently complete to be of value? In today's global competitive world, the authors find it more and more difficult to gain access to useful data about maintenance practices and results, and we aren't even a competitor!

4. But let's say that you manage to solve problems 1, 2, and 3 above. What do you do with it—especially if it shows that you are below par? So you appear to be behind the power curve—how do you fix it? Is there some specific process, procedure, software, or hardware that will do the trick? Or is the difference rooted in factors such as culture, management style, or maybe even the charisma or leadership of a single person? Usually, the benchmark information does not thoroughly answer those questions. We raise the above points because our observation is that many of our clients spend sizeable resources on fairly elaborate benchmarking programs because it is a "valuable exercise." But we see scant evidence to support the notion that benchmarking has ever produced quantum jumps in maintenance optimization. Even companies producing essentially like products (IBM and Apple, Ford and GM, etc.) are so different in their basic business practices as to make benchmarking results rather ineffective in the long-term picture.

There is, however, one facet of benchmarking that can produce benefits (although a complete benchmarking exercise is not necessarily required to gain this benefit). Here we refer to <u>Best Practices</u> which attempt to identify specific actions, processes, and/or equipment that are directly correlated to specific achievements, gains, or efficiencies that may be of value to you. Usually, every organization has at least one Best Practice that others would emulate if they knew about it, even if their overall O&M effort is considered to be below par. In fact, we frequently find that individual plants within the same company have developed certain techniques or procedures that clearly reflect a "Best Practice," but they have failed to pass their expertise on to their fellow organizations (Ref. 6 is a good example that directly deals with RCM). And even when an attempt is made to do so, those at the receiving end often refuse to accept the gratis benefit because "we are not like them" or "they don't understand our problems" or the ever present "not invented here." Referring back to Sec. 1.2, Issue #4, we see that one of the major problems encountered in the maintenance world is the inability to institutionalize

Best Practices. Nevertheless, the potential for transfer of experience that is proven to be the right thing to do suggests that the pursuit of Best Practices is a very worthwhile endeavor.

1.4 MAINTENANCE OPTIMIZATION—AN EMERGING VISION

1.4.1 The Motivating Factor

Whenever we initially meet with a new client, the first (and ultimately most important) point that we introduce at the meeting is the fact that there is only one reason for our discussion—MONEY! No one has ever disagreed with us. But when the client introduces the meeting it is rare, indeed, that their view of the purpose reflects this tone. Rather, they want to jump quickly into the technical issues or the methodology of what to consider. This has frequently led to some rather strange (to us) conclusions when the subject eventually turns to money (if it ever does without our urging). Here are two examples of what we mean.

1. For years, the U.S. electric utility industry has had the mistaken notion that the way to improve its maintenance business is to reduce the cost of preventive maintenance (PM)! No mind that this would eventually increase corrective maintenance actions and reduce plant output to the grid. Until recently, those latter issues were secondary to the reduction of PM, ostensibly because the reduction of scheduled expenditures (i.e., PM) was a primary goal by which a maintenance superintendent was measured. Fortunately, this is changing, along with the management awakening to the fact that they are in a competitive environment where the cost of equipment failure and loss of megawatts could force the customer to other, more cost-conscious competitors.

2. Meetings with clients sooner or later naturally turn to the subject of cost (in our case, RCM program cost). That, of course, is to be expected. But what is almost always missing from their viewpoint is the consideration of return on investment (ROI). Why does this happen? For example, if it is reasonable that a commitment of $40,000 today has a very high probability of eliminating at least just one day of plant outage over the next five years at a saving of $800,000, does that sound like a worthwhile proposition to pursue? That's a minimum ROI of 20:1, and we all tend to bet our personal fortunes with much less likelihood of success (were you one of the dot.com crowd in 2001 who lost their shirt?).

When your initial focus is on MONEY, your perception of O&M improvement changes from reduced PM cost to an increased ROI from your maintenance expenditures. This is a significant, if not dominant, factor in the development of maintenance optimization strategy. The basic reason behind this is the realization that PM costs are quite small in comparison to the price that can ultimately be

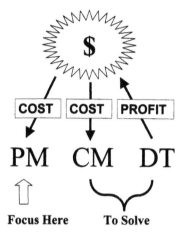

Figure 1.1 Maintenance optimization strategy.

paid if the resulting CM (reactive maintenance) and loss of output (unavailability) are not properly addressed. Figure 1.1 provides a simple but very realistic picture that depicts the model that should drive maintenance optimization strategy.

Notice that two of the money factors in Figure 1.1 are cost items. Most organizations look only at these two elements of this picture—and thus treat maintenance as a cost burden in their company that must be tolerated, but minimized in any way possible. But look what happens if you are smart enough to consider the entire picture, and include the third element of plant OUTPUT. Your first conclusion should be that the financial loss from reduced plant output might likely overshadow any money that might be spent on PM or CM if your output is less than its rated capacity. Your second conclusion should be that the maintenance contribution to loss of output might be a major factor of concern if the record shows that the occurrence of CM events is large and/or shows an increasing trend over the past 12 to 24 months. In other words, maintenance has been and continues to be a major factor in the company's ability to achieve PROFIT! We believe that this suggests a very strong argument for our belief that <u>maintenance is actually a PROFIT CENTER for your company</u>, not just a cost factor to be tolerated as a part of the company's cost of doing business. This may be a very new notion to some of you, but the acceptance of maintenance as a PROFIT CENTER (not Cost Center) is a real element in producing dramatic bottom-line impact, and the first step in moving your organization toward what constitutes World Class Maintenance.

1.4.2 The Traditional Maintenance Mindset

Engineers and technicians are always "optimizing" something. Just about every product that we encounter in the industrial or commercial marketplace is essentially billed as a better mousetrap than its predecessor because some feature is improved (read "optimized") in the new model. We also optimize the various processes, activities, and methods that are employed to design, build, test, and operate these products. Terms such as design optimization are commonly employed in the technical world. In recent years, we have added Maintenance Optimization to our lexicon, and, as management focus has more frequently turned to various O&M issues, we hear more and more about Maintenance Optimization.

Webster defines Optimization as "to make as effective, perfect, or useful as possible." Thus, we could reasonably define Maintenance Optimization as "making our inspection, servicing, and repair/replace actions as effective and useful as possible."

When we look back over the 1950s, 1960s, and 1970s, maintenance engineers and technicians attempted to follow this definition by keeping all equipment within their purview in a serviceable state. And, more often than not, this involved intrusive actions which were both time consuming and susceptible to human error. The result was an increase in return visits or service calls, hardly surprising when there is much evidence that such human error occurs about 50% of the time. As equipment complexity, consumer demand for high levels of quality and availability, and worldwide competition steadily increased, so also did the need develop for more maintenance and maintenance resources (people, tools, software) in order to meet our objective to keep everything in a serviceable state.

Simply put, what had developed as a mindset in the maintenance world was a practice that was based on the notion to *preserve equipment*. That is, keep everything up and running—or at least, available to run and in a perpetual serviceable state.

1.4.3 Rethinking Maintenance Strategy

The 1980s produced a rude awakening across the U.S. industrial complex. What had been essentially a U.S. monopoly in the world marketplace for $3\frac{1}{2}$ decades suddenly became a global arena of fierce competition for virtually every product line that was "made in the USA." Not only did the price of goods to the ultimate consumer become the dominant force, but the bar was continually being raised on consumer demand for quality and service. We all know what happened—in many areas we lost our shirt! The Asian assault had been coming on for many years, swamping one industry after another. In fact, certain industries and products all but vanished from U.S. industry (steel, ships, cameras, home electronics, etc.). With what remained, the battle cry was "Do More With Less." Our vocabulary overworked words such as downsizing, right sizing, re-engineering, continuous improvement, TQM, TPM, and the like.

In all of this, two things happened in the maintenance world which have changed its very soul forever. First, management found that capital investment for new plants and equipment was frequently not a viable option if the bottom line was to remain favorable. What this meant to the maintenance world was that the reliability and availability of existing plants and equipment had to increase. Second, like everyone else, maintenance was also called upon to "Do More With Less." In other words, the resource-intensive philosophy of Preserve Equipment had to change.

As we began to rethink our maintenance strategy, we also began to realize that the Preserve Equipment mindset was seriously flawed—a situation that had been masked for $3\frac{1}{2}$ decades by the monopolies which we enjoyed. For example, four of the most serious problems that went unnoticed for decades were:

1. We had inadvertently promoted an environment where everything was equally important. Every valve, pump, and motor in our plants was maintained to keep it in a serviceable state irrespective of its level of importance to production, safety, or quality.
2. We had also created an environment where decisions were made to perform maintenance simply because there was an opportunity to do so. In a typical scenario, a plant would schedule an outage to specifically service, say, a boiler. But we would then additionally schedule all sorts of maintenance on other equipment just because it was possible to do so. (Do you ever ask your car service shop to take the transmission apart to make sure it is OK just because the car is there to have the brake pads replaced?)
3. As a result of the above, we were overly conservative on our maintenance actions, actually doing more than was necessary in some instances. This often resulted not only in wasted resources, but also in damage to the equipment from our intrusive actions.
4. Then, as the downsizing syndrome set in, the Preserve Equipment modus operandi really overwhelmed us because we couldn't keep up with everything. Our planned maintenance actions fell further and further behind, backlog mounted, and, worst of all, we found our operation becoming almost totally reactive. We were working overtime just to fix the items that were broken.

A new paradigm was needed to address all of these issues. We had to sharpen our decisions on how to effectively use resources that were shrinking, and we had to introduce new technologies that would refine our brute force approach to maintenance. Most importantly, we had to change back from a reactive to a proactive mindset. This new paradigm is embodied in the RCM process.

1.4.4 Focusing Resources—The 80/20 Rule

Manpower resources are diminishing, and will continue to do so. Invoking the new maintenance paradigm is not just another flavor-of-the-month trial balloon. It is a necessity, and is here to stay!

It's obvious, therefore, that our available manpower resources must be applied in a manner that will produce the maximum results. We must focus these resources for the optimum return-on-investment (ROI). The burning question is "how do we do this?".

One answer, and perhaps the best answer among the possible choices, is to apply the 80/20 rule. Simply stated, the 80/20 rule generally asserts that a minority of causes, inputs, or efforts usually lead to a majority of the results, output, or rewards. Historically this imbalance has been observed to be an 80/20 relationship; that is, for example, 80% of the consequences derive from 20% of the possible causes. Here are two typical examples:

- 20% of criminals account for 80% of the value of all crime.
- 20% of a company's products usually account for 80% of their dollar sales value.

An interesting treatise on the 80/20 rule can be found in Ref. 7.

Some of our observed imbalances are 70/30 or 90/10. But the important fact here is that these imbalances do exist, and it is important in our business environment to know where certain of these reside. From a maintenance point of view, the most important imbalance for us to understand is that associated with the expenditure of resources for equipment corrective maintenance (CM), plus loss of productivity (and revenue) from facility downtime (DT), and the source of CM and DT. If the 80/20 rule is valid, then 80% of our CM and DT costs (losses) come from 20% of the systems in our facilities. And we do know that 80/20 is valid from the many applications of this rule that have been measured throughout industry.

Thus, our ability to "preserve" what is really important requires that we know which of our systems are the 20% bad actors. Only then can we focus and allocate our manpower resources properly to make our actions as useful as possible.

Our recommended Maintenance Optimization strategy uses the RCM methodology to focus our resource allocation on those preventive (proactive) maintenance (PM) actions that can reduce or possibly eliminate the CM and DT losses. RCM uses the 80/20 rule as a guide to the application of the Classical RCM process (see Chapter 5) to the "bad actor" systems. That is, the most comprehensive form of RCM is used only where the potential ROI is the greatest.

When dealing with the "well behaved" systems (i.e., the 20/80 systems), we use the Abbreviated Classical RCM™ process (see Sec. 7.2) or the Experience-Centered Maintenance™ process (see Sec. 7.3). In other words, our up-front investment on the 20/80 systems is minimized because the potential gains are commensurately small.

We develop our 80/20 information via the use of Pareto diagrams, which use recent CM and DT cost histories to establish the rank order of facility systems (see example in Chapter 5, Figures 5.1 and 5.2). This information is also tempered by certain mitigating circumstances, if necessary, to avoid the selection of a system that may be driven by a single design problem or is considered an unworthy candidate for PM due to its makeup (such as with digital control systems). By being selective in this fashion, the necessary steps have been taken to assure that the preventive maintenance actions will be directed toward the best possible return-on-investment (ROI).

1.5 WORLD CLASS MAINTENANCE (WCM)—OUR APPROACH

Most discussions of WCM have a rather lengthy list of ingredients that must be in place to qualify as such. The sum of our experiences reflected in the discussions of Secs. 1.1 to 1.4 suggest which key ingredients should be pursued to establish a World Class Maintenance program. There are five simple ingredients in our proposed approach:

1. View maintenance as a PROFIT CENTER.
2. Focus resources for best ROI.
3. Avoid intrusive maintenance.
4. Measure results.
5. Employ an effective management system.

Each of these five ingredients is further reviewed below.

1. *View maintenance as a Profit Center.* This suggests that the maintenance organization must be treated as a key element in your business strategy and plans for achieving profit targets. Maintenance, like any organization function (design, manufacturing, marketing) incurs costs in performing its routine tasks. But recognize that routine (i.e., scheduled) maintenance tasks, when properly performed, have the capability to dramatically affect the ability to achieve or exceed targeted production output. This means, among other things, that Operations (Production) and Maintenance must be treated as equals. No longer should Operations dictate when Maintenance can or cannot do its job. Rather, there must be decisions made for the common good, and each must respect the other's role in meeting the customer demands (i.e., the "real" customer, the one who pays your salaries). No longer should Maintenance feel that its one and only customer is Operations—it is not!

2. *Focus resources for best ROI.* Utilize a structured and systematic process for deciding where you must spend the vast majority of your proactive resources in order to realize the best ROI. After 20 years of research and applications that address the issue, we are convinced beyond any doubt that the Classical RCM process is the correct way to

make those decisions. When properly employed, the RCM process not only identifies where in your plant or facility the bad actor (80/20) systems reside, but also pinpoints exactly where in those systems you must take action to assure that they behave in a reliable fashion. The remainder of this book, of course, is devoted to a complete discourse on the proper use of the Classical RCM process which we suggest is the key to any effort that can be labeled World Class Maintenance.

3. *Avoid intrusive maintenance.* In a separate but related issue, our comprehensive review of current preventive maintenance practices employed by our clients reveals that the majority of current PM tasks involve some form of intrusive action on the part of the craft technicians. These intrusive actions tend to generate return service calls up to 50% of the time due to errors created by the intrusion (see Sec. 2.5—Risk). We believe that a WCM program will employ every possible method and/or technology to reduce intrusive actions to a minimum until it is absolutely necessary to cross the boundary of an equipment item for servicing. This means that maximum use will be made of Condition-Directed tasks including the application of the ever-expanding field of predictive maintenance technology (PdM).

4. *Measure results.* A major problem in most of the current maintenance programs that we have personally witnessed is their inability to effectively collect and use data that describe certain fundamental items of technical and cost information that are necessary to manage and control the maintenance program. Some of the important information items would include the following:

   • Identify those PM tasks that were defined by an RCM study so that their effectiveness in reducing or eliminating CM can be continually evaluated.
   • Collect meaningful equipment history files in order to establish time–failure profiles and as-found equipment conditions that can adjust the intervals assigned to PM tasks, and can initiate root-cause failure studies as may seem appropriate.
   • Perform automated trend analyses from Condition-Directed PM tasks, including automatic preset alarms to warn of an impending approach of critical equipment failure modes.
   • Track the actual trends of maintenance costs and system availability factors in order to measure the overall impact of the maintenance optimization programs—and adjust as required.
   • Utilize these measurements to continually adjust and improve the maintenance program (i.e., the Living RCM Program—see Chapter 10).

5. *Employ an effective management system.* Implement effective management techniques and supporting information systems that will assure the efficient use and control of the critical support and administrative

function of your maintenance organization. This support would include areas such as:

- Work order records
- Inventory control
- Material and manpower usage
- Purchasing and related logistics
- Training
- Craft certifications
- Policies, procedures, and standard instructions
- Scheduling and planning documentation
- etc.

The above information is most commonly acquired and procured via the implementation of a Computerized Maintenance Management System (CMMS).

This book has as its objective a detailed treatment of item #2, RCM. Where appropriate, items #3, #4, and #5 are treated only as they support the proper use and implementation of the RCM process. However, your attention is directed to Appendix C, The Economic Value of Preventive Maintenance, where a detailed discussion is given on the relationship between maintenance and production from a profit point of view. This discussion directly supports item #1 above as a WCM feature.

2

PREVENTIVE MAINTENANCE— DEFINITION AND STRUCTURE

In Chapter 2, we develop several basic elements of a preventive maintenance program. Initially we define preventive maintenance (as distinct from corrective maintenance, which is often a source of confusion), and delineate why preventive maintenance is performed (which many people tend to view too narrowly). This leads then to a discussion of four major PM task categories that can be employed. A logical process for formulating a PM program is suggested, followed by the authors' views and experiences on the current practices and myths that are employed to specify equipment PM tasks. Finally, we briefly examine some of the key disciplines, both management and technical in nature, that can and should be used in supporting PM programs.

2.1 What is Preventive Maintenance?

At first glance, to ask "What is preventive maintenance?" seems to be a rather mundane, if not totally unnecessary, question to pose. However, experience has clearly shown that some confusion does exist over just what people mean when they use the term *preventive maintenance*. There is a variety of possible reasons for this confusion. One significant factor stems from the evidence that a vast majority of our industrial plants and facilities have been operating for extended periods, years in many cases, in a *reactive* maintenance mode. That is to say that the maintenance resources have been almost totally committed to responding to unexpected equipment failures and very little is done in the preventive arena. Corrective, not preventive, maintenance is frequently the operational mode of the day, and this tends to blur how many people view what is preventive and what is corrective. In one actual extreme case, a plant developed an entire culture that fostered a feeling of pride in people's ability to fix things rapidly and under pressure when a forced outage occurred, and rewards were consistently given for such performance. The operating philosophy under these conditions was almost totally reactive and corrective in nature, but plant personnel viewed their actions as preventive in the sense that they were able to "prevent" a long outage because of their highly efficient and effective reactive and corrective actions. What the plant staff did not consciously recognize (or acknowledge) was that they were the highest cost per unit producer among their peers!

Throughout this book, we shall use the following definition of preventive maintenance (PM):

> *Preventive maintenance* is the performance of inspection and/or servicing tasks that have been *preplanned* (i.e., scheduled) for accomplishment at specific points in time to retain the functional capabilities of operating equipment or systems.

The word preplanned is the most important one in the definition; it is the key element in developing a proactive maintenance mode and culture. In fact, this now provides us with a very clear and concise way to define corrective maintenance (CM):

> *Corrective maintenance* is the performance of *unplanned* (i.e., unexpected) maintenance tasks to restore the functional capabilities of failed or malfunctioning equipment or systems.

As viewed by the authors, the entire world of maintenance activity is fully encompassed in these two definitions. It is black and white; there are no gray areas.

However, there are two troubling factors that people frequently question which give rise to some of the confusions over the "PM or CM" discussions. The first of these involves the games that people play with the terminology and its interpretation. These games can be driven by such diverse nontechnical factors as accounting practices or political (regulatory) pressures. For example, some plants, in addition to planned outages for major preventive maintenance tasks and forced outages for unexpected failures with resultant shutdown or cutback in operations, have a third category known as a *maintenance outage* (MO). The MO is historically a measurement of choice used by the generating side of the electric utility industry, and other industries have followed in similar paths. The MO occurs as a result of an *unexpected* equipment problem which hasn't quite yet reached the full failure state but will do so very soon (e.g., within hours or a few days). So the plant management will delay the shutdown or cutback until some off-peak period when the plant outage is more tolerable, and hope that the equipment will hold out until then. Now from an operational point of view, this is a very smart thing to do—but, as a rule, MOs are not counted when it comes to reporting the plant forced outage rate. Somehow they seem to wind up in the preplanned category ("after all, we planned to fix it next Saturday!"). Make no mistake about it, an MO is a forced outage and should be labeled as such when measurements are made. You are only kidding yourself to do otherwise.

A second and more dominant area of confusion occurs when a <u>scheduled</u> task reveals an unacceptable equipment deterioration (like the problem above in the MO situation, except it was not unexpected since a PM task discovered its presence). So actions are taken to repair/restore the full functionality before an unexpected operational impact can occur. Is the repair/restore action preventive or corrective? If you will recall that the purpose of the PM task is to perform actions that will retain functional capabilities, then the answer is essentially self evident—the repair/restore action is <u>preventive</u>. Why? Because a proper structuring of the PM task will <u>always</u> include not only the search for equipment condition, but also the requirement to do something about it if the search uncovers a problem. This search includes PM tasks that require inspection, monitoring parameters that detect failure onset, discovery of hidden failures, and even restoration of equipment that was deliberately allowed to run to failure (see Sec. 2.3 for more details on these items). Unfortunately, though, many CMMS programs will not allow the user to create or code a new work order to cover the emergent work as PM. This additional PM work can only be coded as CM. This is "bean counting" at its worst as it inappropriately inflates the cost of CM, and can lead management to question why CM costs are increasing even when their PM program had been recently improved.

As a general rule, corrective maintenance is more costly than preventive maintenance. As the man in the Pennzoil ad says, "Pay me now or pay me (more) later." And of course we all know the old saying "An ounce of prevention is worth a pound of cure." These catchy phrases are not just idle conversation pieces. They come from the experience of hard knocks. If anyone should doubt this, then just compare two similar plants or systems where one has a proactive maintenance program and the other a reactive maintenance program. Which one do you think has the lower overall maintenance cost and higher availability?

In later chapters, we will find that the use of run-to-failure as a PM task option in a proactive maintenance program is occasionally a very viable option if it is done under very carefully defined and controlled circumstances, which we will spell out in detail in Chapter 5. But the general rule that corrective maintenance should be avoided in favor of preventive maintenance is still the proper way to think when you stop to realize that CM events most often have the consequential effect of also producing a plant or system outage.

2.2 WHY DO PREVENTIVE MAINTENANCE?

This question, too, appears on the surface to be mundane—perhaps even unnecessary. However, for the past 15 years, as part of our seminars and client training programs, we frequently ask the question "Why do preventive maintenance?" and, as a result, we consider it important to raise this question early in the book.

The answers that we consistently hear reflect the popular belief that PM is done for a rather narrowly defined reason and this, as such, leads to the exclusion of a number of golden opportunities for PM enhancement.

So why do you do preventive maintenance? The overwhelming majority of maintenance and plant engineering personnel will respond "To prevent equipment failures." Would that have been your response? If so, you are correct—but not complete in your viewpoint. Unfortunately, we are not yet smart enough to prevent all equipment failures. But that does not mean that our ability to perform meaningful preventive maintenance tasks must end there. In fact, there are three additional and important options to consider. First, while we may not know how to prevent a failure, frequently we do know how to detect the onset of failure. And our knowledge of how to do this is increasing every day, and is creating a whole new discipline called predictive maintenance. Second, even though we may not be able to prevent or detect the onset of failure, we often can check to see if a failure has occurred before an equipment is called into service. Various standby and special purpose equipments (whose operational state is often hidden from the operator's view until it is too late) are candidates for this area. Thus, discovery of hidden failures is yet another PM option available to us. There are also situations in a well planned PM program where economics and/or technical limitations can dictate a decision to do nothing—the appropriately labeled Run-To-Failure (RTF) option. This option, when properly exercised, is done under carefully controlled conditions that are further discussed in Sec. 2.3 and Chapter 5. This RTF option is not to be confused with the more general situation of missing potentially useful PM actions due to oversight or lack of attention to PM planning.

To summarize, there are four basic factors behind the decisions to define and choose preventive maintenance actions:

1. Prevent (or mitigate) failure occurrence.
2. Detect onset of failure.
3. Discover a hidden failure.
4. Do nothing, because of valid limitations.

2.3 PREVENTIVE MAINTENANCE TASK CATEGORIES

By identifying the four factors for doing preventive maintenance, we have also set the stage for defining the four task categories from which a PM action may be specified. These task categories, by one name or another, are universally employed in constructing a PM program, irrespective of the methodology that is used to decide what PM should be done in the program. The four task categories are as follows:

1. Time-directed (TD): aimed directly at failure prevention or retardation.
2. Condition-directed (CD): aimed at detecting the onset of a failure or failure symptom.

3. Failure-finding (FF): aimed at discovering a hidden failure before an operational demand.
4. Run-to-failure (RTF): a deliberate decision to run to failure because the others are not possible or the economics are less favorable.

Each of these will be discussed in more detail to clarify what they cover and how they might be used.

2.3.1 Time-Directed (TD)

In the not too distant past, virtually all preventive maintenance was premised on the basis that equipment could be periodically restored to like-new condition several times before it was necessary to discard it for a new (or improved) item. This premise thus dictated that equipment overhauls were about the only way to do preventive maintenance. Thus at specified "hard time" intervals, overhauls were done regardless of any other consideration. These hard time intervals can be specified in a variety of ways such as clock time, cycles, calendar days, seasons of the year, prior to some defining event, and the like. But the rule is to proceed with the overhaul action when the hard time is reached without any other conditional considerations. Today, we are slowly realizing that this is not always the correct path to pursue. However, in many valid situations we still specify PM tasks at predetermined ("hard time") intervals with the objective of directly preventing or retarding a failure. Conditions under which this approach is valid are discussed in Chapter 5. When such is done, we call it a *time-directed* or TD task. A TD task is still basically an overhaul action—sometimes very complete, extensive, and expensive (like rebuilding an electric motor), and sometimes very simple and cheap (like alignments and oil/filter replacements). As a rule of thumb, whenever we have a planned <u>intrusion</u> into the equipment (even just to inspect it), we have in essence an overhaul-type action which is labeled a TDI (Time-Directed Intrusive) task. Some time-directed tasks can be non-intrusive, such as simple visual inspections or minor adjustments that do not require a breach of the equipment boundary or housing. In this case, the action is simply labeled as a TD task even though the action was performed against a "hard time" interval. But, more often than not, time-directed tasks tend to be intrusive. A simple example that everyone can picture is the changing of oil in our automobile. Here, we intrude in the PM action by removing the drain plug (which will leak if not properly reinstalled), by injecting fresh oil (which must be of the correct type, grade, and quantity with the fill cap properly replaced), and by replacing the oil filter (which will leak if the gasket is not properly installed). The "hard time" associated with this action is car mileage, which has been suggested by the manufacturer who has collected years of experience defining excessive engine wear as a function of oil deterioration due to contaminants and loss of viscosity. Notice that this simple PM task, a TDI task, presents several opportunities for human error to creep into the procedure. And we believe that most of you have personally experienced one or more of these errors at one time or another. Our concern about human error was first mentioned in Sec. 1.2, Item #3, and will be further treated in Sec. 2.5. In fact, we feel so strongly about this potential human error problem that one of our five points associated with World Class Maintenance is to employ every means possible to avoid the use of TDI tasks (see Sec. 1.5). In this illustration of an oil change in automobiles, however, we should also note that this is the correct action to pursue since the relationship between engine wear/failure and contaminated oil is so well established that it would be foolish economics to ignore the facts. Unfortunately, such cause–effect relationships are not usually that well defined.

The keys to categorizing a task as time-directed are: (1) the task action and its periodicity are preset and will occur without any further input when the preset time occurs; (2) the action is known to directly provide failure prevention or retardation benefits; and (3) the task usually requires some form of intrusion into the equipment.

2.3.2 Condition-Directed (CD)

When we do not know how to directly prevent or retard equipment failure—or it is impossible to do so—the next best thing that we can hope to do is to detect its onset and predict the point in time where failure is likely to occur in the future. We do this by measuring some parameter over time where it has been established that the parameter correlates with incipient failure conditions. When such is done, we call it a *condition-directed* or CD task. Thus, a CD task would prewarn us to take action to avoid the full failure event. If the warning comes soon enough, our action can most likely be taken at some favorable timing of our choice. Note that the CD task is dramatically different than the MO situation in that our knowledge of failure onset is a deliberate and preplanned input, as is the action to be taken when the failure onset is detected; the MO is the result of a totally unplanned occurrence. The CD task, like the TD task, has a periodicity for the measurements, but actual preventive actions are not taken until the incipient failure signal is given. The CD task takes two forms: (1) we can measure a performance parameter directly (e.g., temperature, thickness) and correlate its change over time with failure onset; or (2) we can use external or ancillary means to measure equipment status for the same purpose (e.g., oil analysis or vibration monitoring). With the CD task, all such measurements are <u>nonintrusive</u>. The keys to classifying a task as CD are: (1) we can identify a measurable parameter that correlates with failure onset; (2) we can also specify a value of that parameter when action may be taken before full failure occurs; and (3) the task action is nonintrusive with respect to the equipment. Note that if the parameter behaves in a stepwise fashion, as is often the case with digital electronics, it is probably of no use for a CD task.

Let's take a couple of examples to illustrate the CD task. First, we'll look at a rather simple situation that we all encounter at one time or another—the automobile tire. This is an especially interesting example to review because it illustrates several important points. The CD task that is employed here (as a rule somewhat informally) is a performance monitoring of tire tread thickness. We periodically inspect the tires, or the dealer service department does so at predetermined PM shop visits, and when the tread thickness reaches 1/32 inch (or the tell-tale wear strips and thread are of equal height), we get new tires. Notice that tire manufacturers do not recommend an automatic replacement at, say, 25,000 miles (a time-directed task) because they cannot accurately predict the proper mileage for such action due to the many variables influencing wear (not the least of which is driver style and habits). Notice that they do, however, recommend the performance of TDI tasks such as wheel alignment and balance at prescribed fixed intervals in order to help us to get the maximum possible life from the tire. But no amount of PM will prevent eventual tire failure (if left unnoticed) during the useful life of the automobile. With a little added sophistication in the CD task, we can record tread thickness as a function of miles traveled for our specific usage habits, and actually predict when replacement is likely to be needed. If we need to plan for such an investment, as is frequently the case with large truck fleets, this prediction information can prove to be invaluable.

A second, and more technically complex, example might be the use of oil analysis on jet engines, where we measure for chemical and solid contaminants as indicators of wearout and/or incipient failure conditions in hidden parts within the engine; or the use of vibration monitoring sensors on rotating shafts where known limits on shaft movement will be breached when bearing failure onset is developing.

Notice that in the tire example, our knowledge may never be sufficient to specify a TDI task for tire replacement (too many variables with too many people involved). But, in the case of the oil analysis and vibration monitoring, it is very likely that our knowledge of the failure mechanisms and causes involved will someday be well understood—and these CD tasks will be replaced with TDI tasks. As a general rule, it can be said that our current knowledge of failure mechanisms and causes is rather sparse (but we do a lot of guessing anyway); thus, the potential for CD application is large. As our knowledge base increases, we should see the gradual shift from CD to TD or TDI tasks. This shift will be a long-term evolutionary process.

2.3.3 Failure-Finding (FF)

In large complex systems and facilities, there are almost always several equipment items—or possibly a whole subsystem or system—that could experience failure and, in the *normal* course of operation, no one would know that such failure has occurred. We call this situation a *hidden failure*. Backup systems, emergency systems, and infrequently used equipment constitute the major source of potential hidden failures. Clearly, hidden failures are an undesirable situation since they may lead to operational surprises and could then possibly initiate an accident scenario via human error responses. For example, an operator may go to activate a backup system or some dormant function only to find that it is not available and, in the pressure of the moment, fail to take the correct follow-up procedure. So, if we can, we find it most beneficial to exercise a prescheduled option to check and see if all is in proper working order. We call such an option a *failure-finding* (FF) task.

Let's look at a couple of examples to illustrate our point about hidden failures and the FF task. First, look at a simple example—the spare tire in our automobile. If you are like us, you don't really worry about a flat spare tire because you have AAA coverage, and are never more than 10 to 15 minutes away from an ability to get emergency road service—except for that once-a-year trip with the family into "uncharted lands" (e.g., Death Valley). Again, if you are like us, you do check the spare before you leave—and that is a failure-finding (FF) task. Notice that the only intent in such an action is to determine if the spare tire is in working order or not. We are doing nothing to prevent or retard a flat tire (a TD task) or to measure its incipient failure condition (a CD task). It is or is not in working order. And, if it is not in working order, we fix it. That is the essence of what a failure finding task is all about. (Is it OK? If not, fix it.) In this simple example of the spare tire, notice that there are both TD and CD task alternatives that are available to us. As a TD task, we could elect to check the spare tire pressure at preset intervals which we know to be the limits on the tire's capability to hold the required pressure. (Of course, we really do not know this limit, but we could guess at it anyway in an ultraconservative manner—don't laugh, most maintenance engineers make equally wild guesses every day about such failure states.) Or we could run a pressure-sensing line from the tire to a gauge on the dashboard and closely monitor just when tire pressure goes below acceptable limits—i.e., a CD task. Why don't we do one of these tasks? In the case of the TD task, it's really too much trouble given the alternative of the AAA service or the FF task on that once-a-year trip. In the case of the CD task, we don't want to pay for that option, and that is why it is never offered by any current U.S. automobile manufacturer. In other words, convenience and/or cost considerations often drive us to use the FF task in lieu of TD or CD tasks in situations where hidden failures could occur.

A more complex example might involve a standby diesel generator that would be called into service if a grid blackout occurred. One complicating factor here is that we cannot pinpoint when the demand will occur. Thus, we usually go into some form of periodic surveillance task where we start the diesel generator set and bring it to a serviceable power condition to maintain a high confidence in its readiness state. Does this absolutely guarantee that it will successfully perform when a real demand occurs? Not necessarily. However, studies have been conducted which show that the probability of successful performance upon demand can be optimized with the selection of the proper interval for a surveillance (failure-finding) task.

A third example could involve the use of a particular valve during power plant startup. This is a rather common situation; a few valves are opened only for flow alignment during startup and are then returned to the closed position until the next plant startup occurs. In this case, we really don't have to spend money to maintain these valves. Rather, we could opt to specify an FF task, and simply assure ourselves that the valves are in working order a few days before they will be used, or fix them if a problem is discovered.

2.3.4 CD Versus FF—A Distinction

We have come to recognize that there can occasionally be some difficulty in understanding the somewhat subtle distinction between when a PM task can be CD and when it becomes FF. Consider the hypothetical timeline of some equipment's operating history shown here.

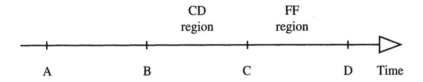

The equipment goes into operation initially at time T_A and runs flawlessly throughout the time interval T_A to T_B. At T_B a failure mechanism that can eventually pose major functional damage or loss is initiated by some type of driving cause. (Please note that in certain instances it is possible that T_B is never reached during the useful lifetime of the equipment item.) But if this failure mechanism continues to develop, it can ultimately reach T_C where the resultant failure mode is reached, and the equipment loses its functional capability. That is, T_C is the time of the equipment failure. If T_C is well beyond T_B, we have a time interval available to us for exercising a CD task if we have the capability to measure a parameter that can track the failure mechanism in question. This can be expressed as follows:

> If $T_C \gg T_B$, CD is possible.

The limiting case here is a situation where $T_B = T_C$ and thus no time is available to perform a CD task. This case can be illustrated by digital electronics which usually perform well in one instance and in the next have completely failed (like your TV set). Also, note the following important fact:

> If $T_C = T_D$, where T_D = time of failure discovery, the failure mode is evident.

Thus, when failure finally occurs, the operators know that something is wrong. But if $T_D > T_C$, then the failure is hidden, and in this interval we have the opportunity to discover the failure (FF).

To summarize, between T_B and T_C, we can possibly define a CD task. Between T_C and T_D we can define a FF task if the failure mode is hidden.

2.3.5 Run-To-Failure (RTF)

As the name implies, we make a deliberate decision to allow an equipment to operate until it fails—and no preventive maintenance of any kind is ever performed. Rather, the maintenance action occurs only after the failure has occurred. There are some limited cases where such a strategy makes common sense, and the details of this strategy will be more fully developed in Chapter 5. Suffice it to say at this juncture that there are three reasons why such a decision can occasionally be made:

1. We can find no PM task that will do any good irrespective of how much money we might be able to spend.
2. The potential PM task that is available is too expensive. It is less costly to fix it when it fails, and there is no safety impact at issue in the RTF decision.
3. The equipment failure, should it occur, is too low on the priority list to warrant attention within the allocated PM budget.

Note the distinction between FF and RTF. With FF the failure is hidden and we do not want to be surprised by its occurrence if the failure should happen. With RTF, we have made a deliberate decision not to be concerned about failure occurrence, be it evident or hidden, and will simply correct the failure at our time of choosing should it occur.

The specifics of how we go about deciding which type of PM task to employ, and examples to illustrate their usage, will be discussed further in Chapters 5 and 6.

2.4 PREVENTIVE MAINTENANCE PROGRAM DEVELOPMENT

Creating a new PM program, or upgrading an existing PM program, involves essentially the same process. We need to (1) determine what we would ideally like to do in the PM program, and (2) take the necessary steps to build that ideal program into our particular infrastructure and put it into action. This process is illustrated in Figure 2.1, and will be subsequently explained in more detail.

Before anything can happen, we must somehow decide what it is that, ideally, we would like to have in place (i.e., the left side of Figure 2.1). That is, we should develop what we believe is the optimum PM program without imposing any restrictions that might otherwise lead us to select "second best" choices among alternatives. For example, we should not limit our selections to PM tasks that only fit the skills of the current maintenance technicians. Later, if greater skills are needed, we may decide that the burden of training or the necessity for hiring more skilled personnel is not an option that is available. But, in the initial formulation of our PM program, we should go for the best possible (ideal) program that can be conceived so that we have clearly displayed the information that management will need to make the required choices (such as to commit to a training program).

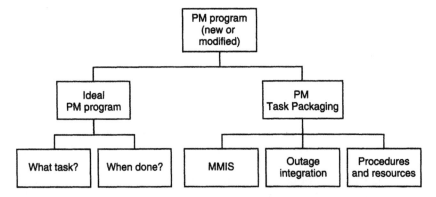

Figure 2.1 Preventive maintenance program development or upgrade.

There are only two pieces of information that are required to define the ideal PM program. Specifically, we must identify (1) *what* PM tasks are to be done and (2) *when* each task should be done. Whatever method may be employed to determine "what tasks," it will result in the definition of a series of tasks that are composed of the task types described in Sec. 2.3. We will briefly discuss some of the "what" methods historically employed throughout industry in Sec. 2.5, and ultimately will recommend the RCM approach in Chapter 4—and for a variety of good reasons. Likewise, methods to determine "when done" information will be covered in Sec. 2.5, and some more specific thoughts on the whole issue of task periodicity will be discussed in Sec. 5.9.

Let us assume for now that we have invoked the recommendations and methodology that are embodied in Chapters 4, 5, and 6, and have defined our ideal PM program. The next job is to incorporate that program, to the maximum extent possible, into the existing corporate infrastructure and assure that it is, in fact, implemented in everyday operations (i.e., integrate the ideal program into the real world, as shown on the right side of Figure 2.1). In order to do this, there is a series of questions and issues that must be resolved before any implementation can occur. Typically, such issues might include the following:

- Are new procedures, or modifications to existing procedures, required?
- Are all of the standard materials available (tools, lubricants, etc.)?
- Is any special tooling or instrumentation required?
- Are any capital improvements required?
- Do we have enough (or too many) people to conduct the program?
- Are the needed skills available? Must training courses be conducted? Is hiring of new skills necessary?
- Does the new/upgraded program affect the spares on-hand (too few, too many)?

- How long will it take to incorporate the new/upgraded program into our CMMS? Is our existing CMMS capable of accepting everything in the new/upgraded program (e.g., tracking time-sequenced data in CD tasks)?
- If we must conduct periodically planned full outages, do the tasks and task intervals lend themselves to such a schedule?
- Do new tasks require a periodicity that is a common denominator with other existing task intervals?

Every situation will have its own unique set of questions which may, or may not, look like the ones in this list. But whatever they may be, it is necessary that we carefully accomplish what is called *Task Packaging*—that is, the specific process for integrating the ideal PM task selections into the existing (or modified) corporate infrastructure for the purpose of putting as many of the ideal PM tasks as possible into the daily operating routine. Only when the ideal PM program is properly integrated with task packaging will we have the operational PM program that we had set about to deploy.

2.5 CURRENT PM DEVELOPMENT PRACTICES AND MYTHS*

As a part of our RCM training seminars, we frequently ask our attendees to describe how their PM program came to be what it is. The responses shed little, if any, light on a solid technical basis of its origin and foundation, and often evoke the candid comment that "I haven't the slightest idea."

With a few notable exceptions (one being commercial aviation), it is a fairly safe bet to say that the vast majority of existing PM programs cannot be traced to their origins; if they can, the origins thus identified still leave open the fundamental question of just why the PM tasks are being done. (Recall that in Sec. 2.1 we noted that some PM programs are, in fact, reactive programs that are more corrective than preventive in nature. These reactive programs, of course, have a particularly difficult time trying to tell you about their origins.) The implication here is that current PM programs are probably wasting resources doing unnecessary tasks or, conversely, are failing to perform necessary tasks, or perhaps are doing some tasks in a very inefficient fashion (e.g., too frequently). The facts tend to support this implication. The various RCM programs that have been conducted repeatedly show hard evidence to the effect that most PM programs meet all of these implications to one degree or another. We will further discuss some of the experience that has been observed in the following paragraphs.

Failure prevention. There is still a widespread feeling in the maintenance community that *all* failures can be prevented. This feeling often motivates the use of

---

*Myth: a notion based more on tradition or convenience than fact (American Heritage Dictionary).

overhaul tasks without any fundamental questioning or understanding about the failure mechanisms involved. As we will see subsequently, and also in Chapters 3 and 4, this overextended use of overhauls can be not only unproductive, but even counterproductive (i.e., create failures that were not present before the overhaul). Some things do wear out and/or deteriorate with age—but probably not to the extent often perceived. We need to identify the failure mechanisms that are involved and, if wearout or aging mechanisms are absent, we should not try to prevent that which is not present initially—it's a waste of money. As our knowledge base becomes more complete, our ability to prevent failures via PM actions will increase. But we need to carefully examine our ability to prevent failure via PM action—before we commit resources that could be misplaced.

Experience. The most common answer given to justify a PM task generally runs like this: "It's been done for 15 years, so it must be good." But did you ever test that hypothesis? Well, ah—no! Let's not mistake for a moment the significant value that resides in experience. But the trick is to use that experience within some logical framework of analysis to lead you to the proper action. The use of "raw" experience is frequently misleading, and perhaps outright wrong.

Judgment. This is the first cousin of experience. It's the extrapolation of experience to a new or (maybe) related area of equipment. If misplaced experience can lead you into trouble, think what misplaced judgment might do! Judgment usually comes in the statement "I think this might be a good thing to do." Rarely said, but almost always implied, is "But I'm not sure I can justify why." One recent endeavor that fits this mold is the use of PM templates. While this approach does attempt to capture a wide spectrum of PM experience and judgment in a structured format, it implies a generalization that "one size fits all." This, in turn, may lead people to neglect the specific tribal knowledge that is required in their plant, and to use template information as "gospel" instead of as a checklist.

Recommendation. This usually comes from the OEM. "The vendor says we ought to do this." The problem with this is that the vendor's recommendations are mainly based on experience and judgment (see previous) and, furthermore, the vendor frequently does not know or understand the specifics of how you will use the equipment. For example, the equipment was designed for steady-state operation, but your application is highly cyclical! Even if the vendor-recommended tasks are correct, they are usually quite conservative on periodicity—especially overhaul intervals. This, of course, might be a good protective measure from their point of view.

Brute force. There seems to be a strong feeling in many quarters that if it is physically possible to do something that appears to have a PM characteristic, then it must be a good thing to do. This is "the more the better" syndrome. It can take some weird forms: overlubrication, cleaning when it shouldn't even be touched, part replacement when there is absolutely nothing wrong with the installed part, etc. Unfortunately, many of these misplaced good intentions are not only a waste of resources, but can also introduce or accelerate equipment failure modes.

Regulation. This is a very difficult area to handle. Most products and services today come under some form of regulatory cognizance—OSHA, EPA, NRC, local PUCs, etc. In their well-meaning ways, these regulators can mandate PM actions that are potentially counterproductive to their objectives. Setting economics aside for the moment, the major difficulty resides in a lack of appreciation of the risk involved in PM actions (see following). By requiring an owner/operator to do certain tasks, the regulators can actually increase the chance for an event (spill, release, etc.) to occur, rather than to help avoid the event. After we educate ourselves, we then need to educate the regulators. Since they are probably here to stay, we should not forget this obligation!

Risk. In this case, the best was saved until the last. There is some conclusive evidence developing that substantiates the "gut feel" of many maintenance engineers that preventive maintenance is, in fact, a potentially risky business. The risk here refers to the potential for creating various types of defects while the PM task is being performed. These defects, or errors, that eventually lead to equipment failures, stem primarily from human errors that are committed during the course of PM task achievement. The risks come in many shapes, sizes, and colors. Typically they may include:

- Damage to an adjacent equipment during a PM task.
- Damage to the equipment receiving the PM task:
  —damage during an intrusion for inspection, repair, or adjustment;
  —installation of a replacement part or material that is defective;
  —misinstallation of a replacement part or material;
  —incorrect reassembly.
- Infant mortality of replaced parts or materials.
- Damage due to an error in reinstallation of an equipment into its original system.

You might wish to reexamine your own records; they are probably replete with evidence of the preceding. And what is especially lethal about this type of "generated" defect is that it usually goes unrecognized—until it causes a forced outage. There has been some published data that illustrates this point—see Figure 2.2 and Ref. 10. This data involved fossil power plants, and examined the frequency and duration of forced outages after a planned or maintenance outage (recall our discussion of MOs in Sec. 2.1). Your attention is drawn to the "Total" column where 56% of forced outages (1772/3146) occurred within one week or less after a planned or maintenance outage! Although these statistics do not reveal exactly how many of those forced outages were due to errors committed during planned outages, there is strong evidence to conclude that over 90% were directly due to errors during the planned outage. Examples include fan blade balance weights knocked off during cleaning (why were they being cleaned?), improper seal seating on overhauled pumps, and missing parts during a reassembly operation. Similar statistics have been observed in other industries. And Ref. 11 describes the

Time \ Duration	<1 week	1 to 2 weeks	2 to 4 weeks	>1 month	Total
<1 week	1705	35	16	16	1772
1 to 2 weeks	358	5	5	2	370
2 to 3 weeks	258	8	0	1	267
3 to 4 weeks	176	0	0	1	177
1 to 2 months	324	12	2	2	340
2 to 3 months	137	3	0	1	141
>3 months	73	3	0	3	79
Total	3031	66	23	26	3146

Figure 2.2 Time between planned or maintenance outage and forced outage versus duration of forced outage.

analysis of problems that occurred during the processing of payloads for the Space Shuttle at Kennedy Space Center where almost 50% were categorized as "generated defects" (i.e., human errors committed in the processing activity). The message is clear—risk is an inherent factor in intrusive actions. One should not perform an intrusive PM task unless he or she is really convinced that there is a justifiable reason for doing it, and then there should be attention to assuring that it is done properly.

If we step back from all of the preceding items, and try to summarize their messages, we could conclude the following: All of them appear to be driven by the principle of "What can be done?" rather than by the principle of "Why should it be done?" The latter is a very key issue and, as we shall see in Chapter 4, was at the source of thinking that ultimately led to RCM.

2.6 PM PROGRAM ELEMENTS

Figure 2.1 gave us a rather simplistic picture of how a PM program is developed. Let's go one step further and look at some of the supporting management and technical disciplines that are involved in the 'Ideal PM Program' and 'PM Task Packaging'. Please note that the disciplines described in this subsection are generally applicable to any PM program, and are not peculiar to any RCM-driven PM program (although they surely do support the RCM concept that will be developed in later chapters).

Ideal PM program. There are a host of supporting technologies that could be listed here. Highlighted are those which we believe are most important. They are

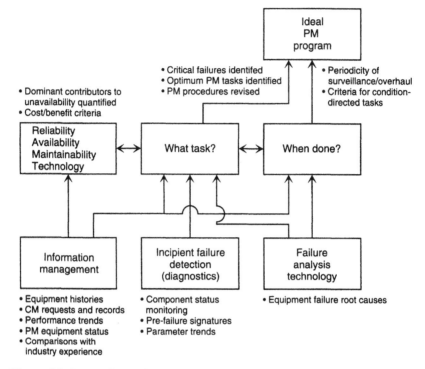

Figure 2.3 Preventive maintenance optimization program.

shown in Figure 2.3, together with a symbolic picture of how they support the "What task?" and "When done?" blocks of Figure 2.1. We will discuss each in more detail.

Failure analysis technology. Consider this important thought. When a design has been completed and then committed to manufacture and use, the designer believes (and hopes) that the product will operate with 100% reliability. In other words, when the design was finalized, the designer had already put his or her best available knowledge into the product. If test and operation of the product prove to be totally successful (a virtually nonoccurring situation), then the designer feels a great deal of satisfaction in the demonstration of the expected product performance. But his or her knowledge base about the product has not been extended—only confirmed. However, if failure occurs, a significant learning opportunity is presented to us. In other words, product malfunctions and failures present us with one of the few important times when we can expand our technical knowledge about the various engineering disciplines—that is, if we take advantage of it! And therein lies the reason for the importance that we give to the conduct of a comprehensive program for failure reporting, root cause analysis, and corrective action feedback. Without such a program, it is virtually impossible to establish the proper correction to the problem, or to intelligently decide if some form of preventive maintenance action is possible.

Let's look at an example to illustrate this point. Consider a motor-operated valve (MOV) that regulates the flow of fluid in a pipe. We have experienced several jamming failures with this valve. We assume (without the benefit of a thorough root cause analysis) that fluid contamination is the culprit. So we (1) install a filter upstream of the valve and (2) tighten the requirements on allowable particulates in the fluid reservoir. Result: the valve still jams. This time we get a little smarter and, performing a thorough (microscopic) analysis, discover that the foreign particles involved in the jamming actions are of the same material as the valve piston! Upon further examination, we also find that the piston design did not properly chamfer the piston circumference, so the edge was breaking off and the particles were wedging between the piston and the cylinder wall. Without this information, we may never have solved the problem, or would have consumed (wasted) valuable resources in a trial-and-error approach. As illustrated in Chapter 3, Figure 3.9, a good failure analysis program is also a vital ingredient in the "retain or increase MTBF" portion of an availability improvement program.

Incipient failure detection. In Sec. 2.3, we discussed the concept of the condition-directed task, and gave several examples of how a CD task might operate. Behind the ability to prudently employ CD tasks is an entire diagnostic technology that is, today, still evolving with new techniques and applications. We believe it is essential to have some form of dedicated effort to follow, understand, and perhaps even contribute to this area that is generally called predictive maintenance technology. This subject is, in a sense, a separate book on its own (see for example Refs. 8 and 9). But, to illustrate its content, listed below are several typical tools that constitute elements of predictive maintenance technology:

- Lubricant analysis
- Vibration, pulse, spike energy measurement
- Acoustic leak detection
- Thermal imaging
- Fiber-optic inspection
- Trace element sensing
- Ultrasonic movement sensing
- Debris analysis
- Creep monitoring
- Dynamic radiography measurement
- Stress/strain/torque measurement
- Hyperbolic moisture detection
- Dye penetrant measurement
- Nonintrusive flow measurement
- Microprocessors with expert system software
- Pattern recognition

Information management. In today's computerized world, it has become necessary to automate the collection, storage, and processing of vital data in order to achieve required levels of operating efficiency. In the operation of large systems, plants, and facilities, such automation is required in the conduct and management of the maintenance program. The Computerized Maintenance Management System (CMMS) is designed to fulfill this need. A typical CMMS will incorporate the following features:

- Automated PM work orders
- PM schedule tracking and measurements
- Corrective maintenance requests and records
- Performance trends
- Failure analysis records
- Condition-directed task measurement, criteria, and alerts
- Equipment history
- Industry equipment experience
- Spares/inventory records
- Skill requirements versus skill availability
- PM and CM cost data

Notice that the CMMS is also a key element in PM task packaging as shown in Figure 2.1.

RAM technology. Reliability/Availability/Maintainability (RAM) technology has a broad spectrum of applicability, and can support a reliability and availability improvement program in many ways. In the area of PM support, RAM models of systems and/or plants can provide the means for predicting and assessing the possible benefits that various PM actions will provide, and for evaluating trade-offs that need to be understood in selecting between competing PM options. As a rule, it is not suggested that RAM models be developed only for the purpose of PM support, since model development can be a costly task when properly done. Rather, if RAM models have been developed as part of a broader application and support to an availability improvement program, they should be used for PM support along with their other uses.

PM task packaging. This subject will be treated more thoroughly in Chapter 8. Here, we would like to indicate briefly three major elements that must be addressed in task packaging (or implementation—carrying PM tasks to the floor).

1. *Task specification.* Recall that the output from the ideal PM program in Figure 2.3 is the "what task" and "when done" information. The task specification is the instrument by which we assure that a complete technical definition and direction is provided to the implementing maintenance organization as to what exactly is required. It is the key transitional document from the ideal to the real world. It is where,

for example, we may first learn of certain constraints that will necessitate a departure from the ideal—and will spell out how this must be handled. As a further example, it will detail the data measurement and evaluation requirements for a CD task along with the limiting acceptance criteria, or will specify critical requirements that must be met in a TDI overhaul task. In some organizations, the task specification is a very formal written process, complete with documentation change control. In other mature organizations, it is rather informal, and frequently is accomplished in meetings that are a prelude to the second element following.

2. *Procedure.* This is the basic document that will guide the field/floor execution of a PM task. In simple PM actions, the procedure may be a one-page instruction, or possibly even a one-line work order authorization. But, in the more complicated PM task, the procedure becomes quite detailed and is considered the "bible" on how the PM task is to be precisely achieved. It should be noted that the risk inherent to PM activities can be controlled and greatly reduced by assuring the development of technically sound and complete task specifications and procedures.

3. *Logistics.* Logistics entails a variety of administrative and production support activities. Typical logistic considerations include tooling, spare parts, vendor support, training, documents and drawings, make/buy decisions (i.e., in-house versus contracted work), test equipment, scheduling, regulatory requirements, etc. Clearly, these considerations closely interplay with both the task specification and the procedure, and constitute a major portion of what is usually called maintenance planning.

In summary, a PM program can be created or upgraded by following the road map of Figure 2.1. The ideal PM program, supported by key technologies (Figure 2.3), will produce the "what task" and "when done" information. In Chapters 4, 5, and 6, we will develop the use of the RCM methodology to supply the "what task" information, and Sec. 5.9 will discuss the "when done" information. This information must then be subjected to the PM task packaging process, described in Chapter 8, to arrive at the specific PM program that can be executed.

3

THE "R" IN RCM—PERTINENT RELIABILITY THEORY AND APPLICATION

Basic reliability concepts play a key role in the underlying philosophy of RCM, and in its implementation. Not everyone, however, is acquainted with the basic concepts of reliability, especially the use of probability and statistics in formulating key reliability principles. This chapter is intended to introduce (or refresh) the reader on certain of these key principles. In particular, the theory portion of the discussion will be done in simplified, qualitative terms with a more mathematically oriented description presented in App. B for those interested in such detail. In addition to the theory aspect, a comprehensive discussion is included on the use of the failure mode and effects analysis (FMEA) technique since this is employed later in the RCM process. Also, the concept of availability, and its role in developing a meaningful maintenance strategy, is discussed.

3.1 INTRODUCTION

Reliability-Centered Maintenance (RCM) has been so named to emphasize the role that reliability theory and practice plays in properly focusing (or centering) preventive maintenance activities on the retention of the equipment's inherent design reliability. As the name implies, then, reliability technology is at the very center of the maintenance philosophy and planning process. It thus seems relevant that we discuss some pertinent aspects of the reliability discipline as a prelude to the specific discussions on RCM that are covered in subsequent chapters.

Our objectives in this chapter are to familiarize the reader with what is commonly called reliability engineering, and then to describe two specific aspects of the discipline that form the application backbone of the RCM methodology: first, basic reliability theory concepts, and second, one of the key reliability tools known as failure mode and effects analysis or FMEA. The theory portion of the discussion will be done in simplified, qualitative terms, but App. B has been included for those interested in a more mathematically oriented description. Refs. 12 and 13 also contain some excellent material for further insights on the subject.

3.2 RELIABILITY AND PROBABILISTIC CONCEPTS

The generally accepted formal definition of reliability is as follows:

> Reliability is the probability that a device will satisfactorily perform a specified function for a specified period of time under given operating conditions.

Thus, satisfactory performance occurs under three specified constraints:

1. Function
2. Time
3. Operating conditions (environment, cyclic, steady state, etc.)

Further, achievement of satisfactory performance is a probabilistic notion, and this introduces the concept of a chance element to the reliability discipline. Satisfactory performance is not a deterministic attribute; it is not the case that it either will or will not happen with absolute certainty. Thus, we must deal with the probability that a device will succeed or fail under specified constraints since it is impossible for us to state with absolute certainty, before the fact, just which outcome will occur.

Consciously or unconsciously, we all deal with probability on a daily basis. For example:

- My car will most probably start without trouble in the morning.
- There is a good chance of afternoon thunderstorms.
- My chance of being involved in an aircraft crash is very small even though I travel frequently by commercial air.
- A single draw from a deck of cards will probably not be an ace.
- There is some finite chance of a second space shuttle accident during the next 100 flights. (With sadness, this finite chance occurred with Columbia in January, 2003.)

- It is highly probable that all of us have made, or heard someone make, one or more of the above statements.

Notice that none of the preceding statements can be made with absolute certainty. However, within the mathematical principles of probability and statistics, it is possible to assign values between 0 and 1.0 to all of these statements if we are willing to work at it hard enough (and spend the money to gather data that is pertinent to each situation). In some instances, it is rather easy to do (e.g., the probability that a single draw will not be an ace is 48/52). In other instances, it becomes difficult to quantify the probability involved, and sophisticated techniques plus costly testing are required (e.g., the probability of a shuttle accident in the next 100 missions). It is indeed rare that we encounter a truly deterministic situation, but some do exist within our current domain of knowledge. For example.

- The sun will rise tomorrow at 6:21 a.m.
- We will die. We will pay taxes.
- Water and oxygen are essential to human survival.

Most of the situations that we encounter in the engineering world have the chance element aspect associated with them. For example, material properties vary, physical environments vary, loads vary, power and signal inputs vary. Some of our basic physical laws can be treated as deterministic—water seeks the lowest accessible level, objects float when they displace their own weight, etc. But when we apply these laws to everyday products, the product performance over time becomes a probabilistic situation. Thus, one would conclude that probability must be a well understood and universally applied discipline in engineering. Unfortunately, not so, although the past decade has seen considerable strides in developing and applying probabilistic design concepts, and the reliability discipline has also made some quantum jumps in probabilistic applications. This latter progression has been motivated by a variety of factors, including issues of safety, regulation, warranty, litigation, and the evolving world competitive market.

So just what are some of the basic aspects that must be considered in the probabilistic sense, and how is this done? Recognize, of course, that complete undergraduate and graduate degrees are given in probability and statistics; anything we can say here must be very abbreviated and simplistic. But a sense of how probability operates will be briefly discussed here.

Most probabilistic events of interest involve counting exercises, ranging from simple to complex. The probability of drawing specified cards from a fair deck illustrates this point well:

$$P\ (\text{ace in a single draw}) = \frac{\text{number of aces in deck}}{\text{number of cards in deck}} = \frac{4}{52}$$

The problem becomes more complicated if we ask for the probability of drawing a jack or a diamond, since the jack of diamonds satisfies both requirements. If we want to know the probability of being dealt a full house in a five-card-draw poker game, added complexities in the counting process are encountered.

Some probability problems involve a question that requires a knowledge of a whole complex population of data. For example, if one should randomly select a person off the street, what is the probability that he or she would be 5'10" or taller? This kind of question introduces the notion of a population height distribution and the necessity to have some reasonable formulation of that distribution in order to answer such a question. We might develop such a distribution for our particular question by examining military service medical records and plotting the distribution of height data therein with the assumption that such is also valid for the whole U.S. population. Such a plot would most likely look like Figure 3.1, which has already been converted to a *probability density function* (pdf). This particular pdf is the familiar bell-shaped or normal distribution which happens to be the distribution that we find for many population characteristics of interest. In Figure 3.2, we find the answer to our question in the shaded area. Since the area under the entire curve in Figure 3.1 is equal to 1.0, then one-half of the area in Figure 3.2 represents our answer, or 0.5. If our question had been directed to the probability that

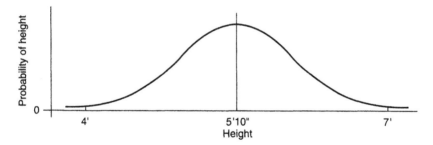

Figure 3.1 The gaussian (normal) pdf for population height.

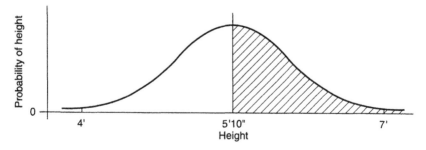

Figure 3.2 Calculating probability (height ≥ 5'10").

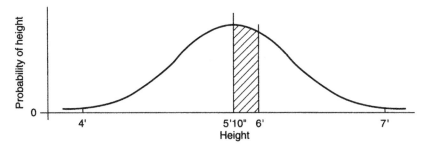

Figure 3.3 Calculating probability (6′0″ ≥ height ≥ 5′10″).

he or she would be between 5′10″ and 6′0″, then Figure 3.3 would provide the answer by calculating the area in the crosshatched section shown. All of this can be done with mathematical precision once the basic pdf for the parameter in question (in this case, height of the U.S. population) is known.

A variety of pdfs can be used to describe the probability of an event of interest. For example:

- The probability of exactly *X* heads in *Y* flips of a fair coin will use the binomial distribution as the basis to describe a population of trials where each trial has only two possible outcomes. Clearly, a single flip of a fair coin is 50/50 for a head or tail. But to calculate the probability of 3 heads in 10 flips of a fair coin before any coin flipping has occurred requires more sophisticated calculations.
- If we know the average rate at which phone calls come into a switchboard, we can use a Poisson distribution to calculate the probability that the switchboard will receive 0, 1, 2, 3, or *N* calls during a 60-minute period. Such information is very useful in decisions on staffing levels or training/skill requirements for hiring switchboard operators.

Several different distributions or pdfs exist, and they tend to be used for different kinds of populations and events that can be described. In Sec. 3.4, one specific distribution, the exponential, will be of particular interest in our discussion of reliability theory.

The point to all of this, again, is that reliability is a probabilistic concept, and thus some basic understanding and appreciation of probability is very much in order.

3.3 RELIABILITY IN PRACTICE

It is basic to the notion of reliability that we have some appreciation of just how this probabilistic aspect in the real world might affect the products that we design, build, and operate. Figure 3.4 is a picture that displays this reality. Suppose we have a product or system composed of N identifiable elements or devices (Figure 3.4 shows cases for $N = 10$, 50, 100, and 400). None of those elements is "perfect," hence the curve presents cases (on the x-axis) for individual element reliability values between 100 and 97 percent (which, as individual element values, are considered very good in many systems). The system requires that all N elements perform satisfactorily for system success. The probability of system success— i.e., the value of system reliability—is shown on the y-axis. Two things are quite apparent in this display:

1. As system complexity increases (i.e., as N increases), system relia- bility drops dramatically even for average element reliability values greater than 90 percent. Thus, if a system must contain a large number of individually required elements, these elements must, per se, have very high reliabilities, or we can expect the system to fail frequently.
2. Even if the system is relatively simple (say, $N = 50$), the individual average element reliability needs only to drop slightly for system reliability to drop significantly!

Thus, the need to concentrate on product reliability is not just a PR game—it is very real in terms of ultimately providing the expected customer satisfaction.

As you might already have noticed, one aspect of any high-reliability product strategy is based on the two items just mentioned—that is, (1) keep it simple and (2) have very high individual element reliability. A third item in many complex products is the use of various redundant techniques.

The important question thus becomes "How do we achieve high system reliabi- lity, and what are the key ingredients that must be addressed?" First and foremost, it must be recognized that reliability is a design attribute. By this, we mean that product reliability is established by how well (or poorly) the design process is accomplished. Reliability cannot be fabricated, tested, or inspected into a prod- uct. The design or, more broadly, the product definition (which also encompasses how the product must be operated and maintained) is the sole determinant in set- ting the inherent, or upper level, of reliability that can be achieved. Fabrication, assembly, test, operation, and even maintenance can only degrade the inherent reliability if they are not performed properly—but none of these activities can enhance it beyond the capability established by the basic design and product definition.

Programs which are frequently organized under the title of "Reliability Engineering" are often employed to bring together a variety of technical and management functions that will concentrate on guiding and assisting the basic engineering functions in achieving the expected product reliability performance.

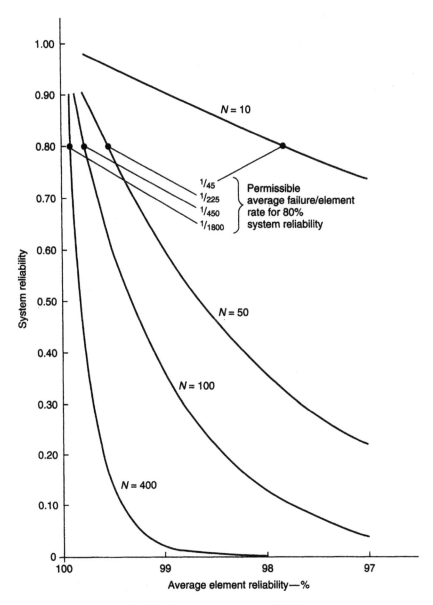

Figure 3.4 Element reliability impact on system reliability.

Some of these reliability engineering functions typically include the following activities:

- Comprehensive review of product specifications to assure that all reliability objectives and supporting requirements are properly included.
- In-depth reliability analyses of the design concept and detail via reliability prediction models, design trade studies, failure mode and effects analysis, critical function analysis, design life analysis, proper use of design standards, incorporation of redundancy and other forms of design margin, etc.
- Continuous review and control of potential risk contributors to the product—e.g., adherence to proper part and material applications, establishment of proper manufacturing and quality control processes, structuring meaningful product test programs, control of design changes, and assurance that failures and problems are thoroughly analyzed and fed back to the design for necessary corrective actions.

It is during the product design and development phase that preventive maintenance tasks are initially specified. The RCM methodology described in subsequent chapters is a highly effective method for developing these initial PM task specifications. Clearly, the design process should recognize the importance of a proper PM program in retaining the inherent reliability of a product. PM actions ranging from simple lubrication tasks to more complex replacement of certain life-limited parts are necessary ingredients in the retention of inherent reliability. Unfortunately, this aspect of the design process is often relegated to secondary priority, and products are then fielded with a less than adequate PM program— and thus, a less than reasonable probability that they will operate up to the customer expectations for reliability. Many products and systems in operation today fall into this category. However, it is still possible to apply the RCM methodology to these products and systems, thereby upgrading their PM programs and ultimately realizing the full potential of the inherent design reliability.

3.4 SOME KEY ELEMENTS OF RELIABILITY THEORY

In Sec. 3.2, we saw that probability calculations derive from a counting process, and that this process may require a knowledge of population data that will describe the parameter of interest. In reliability, the population that will enable us to calculate reliability values is the failure versus time data. In other words, we need to understand the time distribution for how a large number of devices will fail (or die). If we can accumulate enough data to define or approximate such a distribution, we can define a population density function of the failures—or, in this case, a failure density function (fdf). In the mathematics of reliability, the fdf is usually designated as $f(t)$. Once we know $f(t)$, we can calculate reliability, unreliability, and two very important parameters called the death rate and the mortality rate. These latter two terms derive from actuarial statistics which are employed in the insurance business in order to set policy premium payments.

The death rate is defined as the death (or failure) frequency with respect to the original population, while the mortality rate is the failure frequency with respect to the surviving population at some time of interest. A typical life insurance example can illustrate this distinction quite easily. Suppose we look at 1 million people born in 1929, and the records tell us that 10,000 of these people died in 1989. The death rate in 1989 for people born in 1929 is:

$$10,000/1,000,000 = 1/100$$

Now, if only 200,000 of the original 1 million are living on January 1, 1989, then the mortality rate in 1989 for people 60 years old is:

$$10,000/200,000 = 1/20$$

Obviously, insurance premiums are established on the basis of the mortality rate which is usually labeled as $h(t)$ or just λ, and is commonly referred to as the instantaneous failure rate, or just failure rate.

In reliability problems, λ is the parameter of interest to us. In other words, we want to know the probability that devices currently in operation will continue to operate satisfactorily for the next T hours. Or, conversely, what is their probability of failure? If we know the fdf or $f(t)$ for the device, we can calculate all of these values.

In reliability, there is one fdf of special interest that is called the exponential fdf. This special interest arises for two reasons:

1. There is some substantial evidence that many devices (especially electronics) follow the exponential fdf law.
2. Mathematically, the exponential fdf is the easiest to handle. Because of this feature, we often assume that some product, system, or device follows the exponential law—only to find later that such is not true, and we have thus miscalculated the product reliability.

The specific feature that makes the exponential fdf so easy to handle is that the mortality or failure rate, λ, is a constant over time (rather than varying with time). This mathematical nicety means that, in the hardware world, device or product failure is a random process which occurs, on average over an extended period, at some fixed time interval. Stated differently, if λ is a constant, then the failures are independent of time, and will neither increase nor decrease in frequency as the product or device population ages. This feature has some profound implications on preventive maintenance, and these are discussed in Sec. 4.2. Further, the reciprocal of λ, $1/\lambda$, is the mean of the exponential fdf, and is called the *mean time between failure* or MTBF. All pdfs (or fdfs) have a mean value called the *mean time to failure* or MTTF. With the exponential fdf, MTTF = MTBF, but only with the exponential is the MTBF a constant over time. With other fdfs, MTTF is a single value that occurs only once for the distribution represented. Thus, if we do not have an exponential law governing the failure history for a device or product, we will experience different failure rates depending upon where we are in the device or product life cycle. In these instances, it is important to know the details of the failure mechanisms and their causes so that proper design, maintenance, or operation actions can be taken to achieve the specified reliability.

There is a generally accepted concept in reliability that attempts to put both the constant and nonconstant λs together to describe a typical device or product life cycle. It is called the *bathtub curve* (see Figure 3.5). The name clearly derives from its shape, which evolves from the following three scenarios:

1. In the early stages of product deployment, there is some residual of substandard parts, materials, processes, and workmanship that escapes the factory test and checkout actions, and thus remains in the product at its point of initial use and operation. These substandard items generally surface rather quickly relative to the total product lifetime, but initially they produce a failure rate that is larger than the expected long-term failure rate. As these problems surface and are removed, the population failure rate will decrease and a stabilization of the population λ will occur. This first phase of the cycle is called the *infant mortality stage*.
2. When the population stabilization is complete, the constant failure rate phase described previously takes over. Product failures are more random in nature, and we have stabilized at the level of inherent reliability of the product. The product population, *on average*, has a constant MTBF, but because of the randomness in failure occurrence we can predict neither the precise time nor the exact nature of the failures that will ultimately occur.
3. As the product operating life progresses, several potential failure mechanisms may develop which are no longer random in nature. In fact, they are

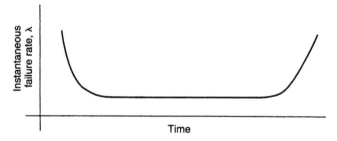

Figure 3.5 The reliability "bathtub curve."

very time- or cycle-dependent, and lead to product aging and wearout. These mechanisms include items such as material wear, fatigue deterioration, grain structure changes, and material property changes. When this happens, the population failure rate will again start to increase, and we see that the product may be nearing its end of useful life if the repair or replacement of the affected parts or devices is extensive and costly. This third phase of the scenario is called the *aging and wearout stage*.

Whether or not an item of equipment has the infant mortality and/or aging and wearout stages likewise has a profound impact on preventive maintenance strategy, and this is discussed in Sec. 4.2. Often, the case may be that infant mortality and/or aging and wearout are known, with a reasonable engineering confidence, to exist, but the times at which these stages occur in the product lifetime are not well defined. In this situation, we also have some decisions to make regarding the choice of a PM task and its periodicity. More is said about this in later discussions, including the discussion on age exploration in Sec. 5.9.

In summary, the key elements of reliability theory that are germane to the RCM methodology are as follows:

1. Knowledge of a product or device fdf allows the calculations of the reliability parameters that may be of interest.
2. A key parameter in this regard is the failure rate, λ.
3. One specific fdf frequently quoted and employed is the exponential fdf wherein λ = constant and is therefore independent of time. (Often the exponential is assumed when, in reality, it is not the proper fdf.)
4. There is a generally accepted depiction of a typical product life cycle known as the bathtub curve.
5. Whether a product or device follows the bathtub curve can have a profound impact on the proper selection of PM tasks.

3.5 FAILURE MODE AND EFFECTS ANALYSIS (FMEA)

The FMEA is generally recognized as the most fundamental tool employed in reliability engineering. Because of its practical, qualitative approach, it is also the most widely understood and applied form of reliability analysis that we encounter throughout industry. Additionally, the FMEA forms the headwaters for virtually all subsequent reliability analyses and assessments because it forces an organization to systematically evaluate equipment and system weaknesses, and their interrelationships that can lead to product unreliability.

But before we proceed to discuss the FMEA process, we feel it is important to address a semantics issue that often arises in this discussion. To put it most succinctly, failure to define failure can lead to some unfortunate misunderstandings.

For as long as we can recall, there have been varying degrees of confusion about what people mean when they use terminology that involves the word "failure." Failure is an unpleasant word, and we often use substitute words such as anomaly, defect, discrepancy, irregularity, etc., because they tend to sound less threatening or less severe.

The spectrum of interpretations for failure runs from negligible glitch to catastrophe. Might we suggest that the meaning is really quite simple:

> Failure is the inability of a piece of equipment, a system, or a plant to meet its expected performance.

This expectation is always spelled out in a specification in our engineering world and, when properly written, leaves no doubt as to exactly where the limits of satisfactory performance reside. So, failure is the inability to meet specifications. Simple enough, we believe, to avoid much of the initial confusion.

Additionally, there are several important and frequently used phrases that include the word failure: failure symptom, failure mode, failure cause, and failure effect.

Failure symptom. This is a <u>tell-tale indicator</u> that alerts us (usually the operator) to the fact that a failure is about to exist. Our senses or instruments are the primary source of such indication. Failure symptoms may or may not tell us exactly where the pending failure is located or how close to the full failure condition we might be. In many cases, there is no failure symptom (or warning) at all. Once the failure has occurred, any indication of its presence is no longer a symptom—we now observe its effect.

Failure mode. This is a brief description of <u>what is wrong</u>. It is extremely important for us to understand this simple definition because, in the maintenance world, it is the failure mode that we try to prevent, or, failing that, what we have to physically fix. There are hundreds of simple words that we use to develop appropriate failure mode descriptions: jammed, worn, frayed, cracked, bent, nicked, leaking, clogged, sheared, scored, ruptured, eroded, shorted, split, open, torn, and so forth. The main confusion here is clearly to distinguish between failure mode and failure cause—and understanding that failure mode is what we need to prevent or fix.

Failure cause. This is a brief word description of <u>why it went wrong</u>. Failure cause is often very difficult to fully diagnose or hypothesize. If we wish to attempt a permanent prevention of the failure mode, we usually need to understand its cause (thus the term, root cause failure analysis). Even though we may know the cause, we may not be able to totally prevent the failure mode—or it may cost too much to pursue such a path.

As a simple illustration, a gate valve jams "closed" (failure mode), but why did this happen? Let's say that this valve sits in a very humid environment—so "humidity-induced corrosion" is the failure cause. We could opt to replace the valve with a high-grade stainless steel model that would resist (perhaps stop) the corrosion (a design fix), or, from a maintenance point of view, we could periodically lubricate and operate the valve to mitigate the corrosive effect, but there is nothing we can do to eliminate the natural humid environment. Thus, PM tasks cannot fix the cause—they can address only the mode. This is an important distinction to make, and many people do not clearly understand this distinction.

Failure effect. Finally, we briefly describe the <u>consequence of the failure mode</u> should it occur. To be complete, this is usually done at three levels of assembly—local, system, and plant. In describing the effect in this fashion, we clearly see the buildup of consequences. With our jammed gate valve, the local effect at the valve is "stops all flow." At the system level, "no fluid passes on to the next step in the process." And finally, at the plant level, "product production ceases (downtime) until the valve can be restored to operation."

Thus, without a clear understanding of failure terminology, reliability analyses not only become confusing but also can lead to decisions that are incorrect.

The FMEA embodies a process that is intended to identify equipment failure modes, their causes, and finally the effects that might result should these failure modes occur during product operation. Traditionally, the FMEA is thought of as a design tool whereby it is used extensively to assure a recognition and understanding of the weaknesses (i.e., failure modes) that are inherent to a given design in both its concept and detailed formulation. Armed with such information, design and management personnel are better prepared to determine what, if anything, could and should be done to avoid or mitigate the failure modes. This information also provides the basic input to a well-structured reliability model that can be used to predict and measure product reliability performance against specified targets and requirements.

The delineation of PM tasks is also based on a knowledge of equipment failure modes and their causes. It is at this level of definition that we must identify the proper PM actions that can prevent, mitigate, or detect onset of a failure condition. Specifying PM tasks without a good understanding of failure mode and cause information is, at best, nothing more than a guessing game. Hence, the FMEA will play a vital role in the RCM process, and this will be developed in more detail in Chapter 5.

How do we perform the FMEA? First, it should be clear by now that a fairly good understanding of the equipment design and operation is an essential starting point. The FMEA process itself then proceeds in an orderly fashion to qualitatively consider the ways in which the individual parts or assemblies in the equipment can fail.

These are the failure modes that we wish to list, and are physical states in which the equipment could be found. For example, a switch can be in a state where it cannot open or close. The failure modes thus describe necessary states within functions of the device which have been lost. Alternatively, when sufficient knowledge or detail is available, failure modes may be described in more specific terminology—such as "latch jammed" or "actuating spring broken." Clearly, the more precise the failure mode description, the more understanding we have for deciding how it may be eliminated, mitigated, or accommodated. Although it may be difficult to accurately assess, we also attempt to define a credible failure cause for every failure mode (maybe more than one if deemed appropriate to do so). For example, the failure mode "latch jammed" could be caused by contamination (dirt), and the "broken spring" could be the result of a material–load incompatibility (a poor design) or cyclic fatigue (an end-of-life situation).

Each failure mode is then evaluated for its effect. This is usually done by considering not only its local effect on the device directly involved, but also its effect at the next higher level of assembly (say, subsystem) and, finally, at the top level of assembly or product level (say, system or plant). It is usually most convenient to define two or three levels of assembly at which the failure effect will be evaluated in order to gain a full understanding of just how significant the failure mode might be if it should occur. In this way, the analyst gains a bottoms-up view of what devices and failure modes are important to the functional objectives of the overall system or product. A typical FMEA format is shown on Figure 3.6.

By way of example, an FMEA is shown on Figure 3.7 which is based on the simple lighting circuit schematic shown in Figure 3.8. In this instance, the FMEA is conducted at the system level due to its simplicity, and we just move around the system circuit, device by device. In a more complex analysis, we might devote an entire FMEA to just one device, and break it into its major parts and assemblies for analysis. A pump or transformer are examples of where this might be done.

Frequently, FMEAs are extended to include other information for each failure mode—especially when the FMEA is conducted in support of a design effort. These additional items of information could include:

- failure symptoms
- failure detection and isolation steps
- failure mechanisms data (i.e., microscopic data on the failure mode and/or failure cause)
- failure rate data on the failure mode (not always available with the required accuracy)
- recommended corrective/mitigation actions

When a well-executed FMEA is accomplished, a wealth of useful information is generated to assist in achieving the expected product reliability.

EQUIPMENT		FAILURE MODE	FAILURE CAUSE	FAILURE EFFECTS		
I.D. #	DESCRIPTION			LOCAL	SYSTEM	UNIT

Figure 3.6 Failure mode and effects analysis.

Component	Mode	Effect	Comment
1. Switch A1	1.1 Fails open 1.2 Fails closed	1.1 System fails 1.2 None	1.1 Cannot turn on light. 1.2 If A2 also fails closed, then system fails by premature battery depletion.
2. Switch A2	(same as A1)	(same as A1)	(same as A1)
3. Light Bulb C	3.1 Open filament 3.2 Shorted base	3.1 System fails 3.2 System fails; possible fire hazard	3.1 Cannot turn on light. 3.2 Cannot turn on light. May cause secondary damage to rest of system.
4. Battery B	4.1 Low charge	4.1 System degraded; dim light bulb	4.1 May be precursor to "no charge."
	4.2 No charge 4.3 Over-voltage charge	4.2 System fails 4.3 System fails by secondary damage to Light Bulb C	4.2 Cannot turn on light. 4.3 Secondary damage to Light Bulb C caused by over-current.

Figure 3.7 Simple FMEA.

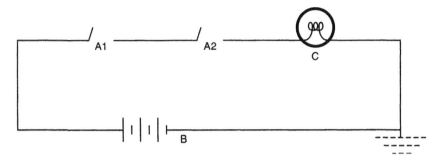

Figure 3.8 Simple circuit schematic.

3.6 AVAILABILITY AND PREVENTIVE MAINTENANCE

In Sec. 1.4 and Figure 1.1 we presented our view that with World Class Maintenance, the organization should be treated as a Profit Center for your company. This view derives from our belief that the overwhelming financial aspects of a maintenance optimization strategy are driven by considerations of downtime (DT) elimination (a.k.a. increased availability)—which, when properly accomplished, is a direct contributor to company revenue and profit. And preventive maintenance plays a major role in this scenario. This section will further discuss this scenario by developing the relationship between availability and preventive maintenance.

First, let's clarify some terminology. We frequently hear discussions about plant productivity that will interchangeably use the terms reliability and availability when referring to technical disciplines that must be invoked. In Sec. 3.3, the broadly accepted definition of reliability was given. In contrast, availability is defined as follows:

> Availability is a measure of the percentage (or fraction) of time that a plant is capable of producing its end product at some specified acceptable level.

Thus, by definition, availability must account for plant outages—both planned (scheduled) and unplanned (forced). Scheduled outages are factored into production commitments as a matter of course, as are forced outages at some small and acceptable level.

To understand how the issue of availability really affects us, consider a hypothetical situation at the ABC Corporation where management has formulated a top-level policy to "maintain output." This is certainly a sound objective for ABC to pursue, but how do we translate this into meaningful tasks for ABC's various plants throughout the world? First, we need to realize that this policy is directly addressing a need to maintain or perhaps even increase availability at its plants. The most direct way to do this is to avoid outages, especially lengthy forced outages. In fact, it is quite common to see specific goals assigned to plants to keep their "forced outage rate" below some specified annual value such as 3.5 percent. If the annual scheduled outage rate is 1.5 percent, then it would be expected that the plant would be capable of producing 95% of the time during the year.

But, down at the plant level, how does the General Manager and his supervisors define specific tasks that will help them to meet or exceed those goals? To answer this question, we need to understand more precisely just what constitutes the availability measure. There are two, and only two, parameters that control this measure:

- Mean time between failure, or MTBF, which is a measure of how long, on average, a plant (or an individual item of equipment) will perform as specified before an unplanned failure will occur.
- Mean time to restore, or MTTR, which is a measure of how long, on average, it will take to bring the plant or equipment item back to normal serviceability when it does fail.

MTBF, then, is a measure of the plant or equipment reliability (R) and MTTR is a measure of its maintainability (M). Mathematically, we can define availability (A) as follows:

$$A = \frac{\text{MTBF}}{\text{MTBF} + \text{MTTR}}$$

Notice that if MTBF is very large with respect to MTTR—that is, if we have a very high plant reliability—availability will also be high, simply because the MTBF parameter dominates what is physically occurring. Conversely, a very small MTTR can also yield a high availability because, even if the equipment fails frequently, it can be restored to service very quickly. Usually, neither of these two limiting cases exist, and we have to work diligently at retaining or improving both the MTBF and MTTR parameters in order to achieve a high degree of plant availability.

Recognizing that MTBF (or reliability) and MTTR (or maintainability) are the parameters that we must influence will now simplify our job considerably. There are several tasks that can be performed, and usually some investigation and evaluation of plant problems and operating practices will reveal where the resources should be focused. In particular, however, the role that an effective preventive maintenance program (PMP) can play in achieving desired levels of availability is of special note. This is true because the PMP can beneficially impact both reliability and maintainability when it is properly specified and conducted. The right preventive maintenance (PM) tasks can, for example, be the primary factor in keeping an item of equipment in top running order—tasks as simple as

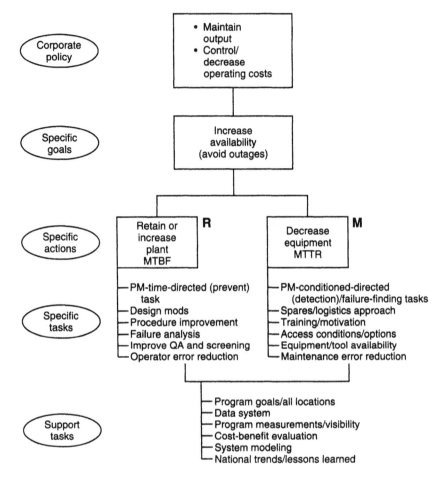

Figure 3.9 Availability improvement program—key issues.

lubrication and alignment checks performed at specified intervals can be the necessary link to retaining an inherent design reliability. In like manner, the right PM tasks could play a major role in decreasing MTTR simply by the use of periodic on-condition monitoring that would detect failure onset and permit an opportunity for repair or replacement at the timing of your choice, thus avoiding the forced outage.

We can summarize our discussion for the ABC Corporation in a single picture shown in Figure 3.9. The PMP is particularly potent because, when done properly, it produces a double-barreled effect by impacting both reliability (MTBF) and maintainability (MTTR) within the same execution.

4

RCM—A PROVEN APPROACH

In this chapter, we will introduce the basic concepts that constitute what is known as Reliability-Centered Maintenance. Initially, however, we will briefly discuss how PM has evolved in the industrial world, and, most importantly, we will look at how one of the basic tenets of reliability engineering—the "bathtub curve"—can and should influence the formulation of PM tasks. Next, we will look at how the commercial aviation industry was historically the motivating force behind the creation of the RCM methodology during the Type Certification process for the 747 aircraft in the 1960s. Finally, we will itemize the four basic features that constitute the necessary and sufficient conditions or principles that define RCM, and discuss some of the cost–benefit considerations that can accrue through the use of RCM.

4.1 Some Historical Background

If we look back to the days of the Industrial Revolution, we find that the designers of the new industrial equipment were also the builders and operators of that equipment. At the very least, they had a close relationship with the hardware that evolved from their creative genius, and as a result they truly did "know" their equipment—what worked, how well, and for how long; what broke, how to fix it, and, yes, how to take certain reasonable (not too expensive) actions to prevent it from breaking. In the beginning, then, experience did in fact play the major role in formulating PM actions. And, most importantly, these experience-based actions derived from those people who had not just maintenance experience, but also design, fabrication, and operation knowledge. Within the limits of then available technology, these engineers were usually correct in their PM decisions.

As industry and technology became more sophisticated, corporations organized for greater efficiency and productivity. This, of course, was necessary and led to numerous advantages that ultimately gave us the high-volume production capability that swept us into the twentieth century. But some disadvantages occurred also. One of these was the separation of the design, build, and operate roles into distinct organizational entities where virtually no one individual would have the luxury of personally experiencing the entire gamut of a product cycle. Thus, the derivation of PM actions from experience began to lose some of its expertise.

Not to worry! Another technology came along to help us—reliability engineering. The early roots of reliability engineering trace back to the 1940s and 1950s. Much of its origin resides in the early work with electronic populations where it was found that early failures (or infant mortalities) occurred for some period of time at a high but decreasing rate until the population would settle into a long period of constant failure rate. It was also observed that some devices (e.g., tubes) would finally reach some point in their operating life where the failure rate would again sharply increase, and aging or wearout mechanisms would start to quickly kill off the surviving population. (This scenario, of course, also very accurately describes age–reliability characteristics of the human population.) Engineers, especially in the nonelectronic world, were quick to pick up on this finding and to use it as a basis for developing a maintenance strategy. The picture we have just described is the well-known *bathtub curve*. Its characteristic shape (seen previously in Sec. 3.4, Figure 3.5) led the maintenance engineer to conclude that the vast majority of the PM actions should be directed to overhauls where the equipment would be restored to like-new condition before it progressed too far into the wearout regime.

Thus, until the early 1960s, we saw equipment preventive maintenance based in large measure on the concept that the equipment followed the bathtub shape, and that overhaul at some point near the initiation of the increasing failure-rate region was the right thing to do.

Some additional historical perspective on the evolution of reliability engineering can be found in Refs. 14 and 15.

4.2 THE BATHTUB CURVE FALLACY

As this title suggests, all may not be totally well with the bathtub curve. True, some devices *may* follow its general shape, but the fact is that more has been assumed along those lines than has actually been measured and proven to be the case. As those with even a cursory knowledge of statistics and reliability theory can attest, this is not surprising, because large sample sizes are required in order to accurately develop the population age–reliability characteristics of any given device, component, or system. And such large samples, with recorded data on operating times and failures, are hard to come by.

The commercial aviation industry, however, does have fairly large populations of identical or similar components in its aircraft fleets—components that are common to several aircraft types. And, as an industry, it has made some deliberate and successful efforts to accumulate a database of operating history on those components. Such a database is driven by several factors, not the least of which are safety and logistics considerations. As a part of the extensive investigation that was conducted in the late 1960s as a prelude to the RCM methodology, United Airlines used this database to develop the age–reliability patterns for the nonstructural components in their fleet. This was done as a part of the more general questioning that preceded RCM concerning whether airline equipments did, in fact, follow the bathtub curve. Specifically, failure density distributions were developed from the component operating history files, and the hazard rate (or instantaneous failure rate) was derived as a function of time. The results of this analysis are summarized in Figure 4.1 (from Ref. 16).

These results came as a surprise to almost everyone—and continue to do so today when people see these results for the first time. Follow up studies using aircraft data in Sweden in 1973, and by the U.S. Navy in 1983, produced similar results, as shown in Figure 4.2.

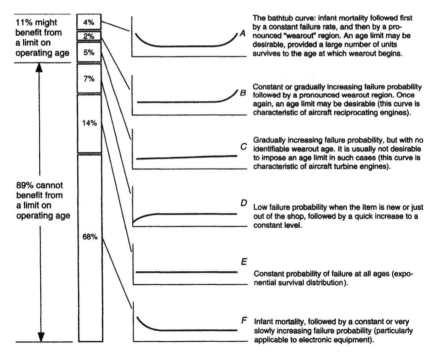

Figure 4.1 Age–reliability patterns for nonstructural equipment (United Airlines).

Failure Rate by Type (from Fig. 4.1)	UAL 1968 (Fig. 4.1)	Bromberg 1973 (Ref. 17)	U.S. Navy 1982 (Ref. 17)
A	4%	3%	3%
B	2%	1%	17%
C	5%	4%	3%
D	7%	11%	6%
E	14%	15%	42%
F	68%	66%	29%

Figure 4.2 Age–reliability patterns.

The Bromberg and U.S. Navy results were extracted from Ref. 17. The United Airlines and Bromberg results are essentially identical findings, and the U.S. Navy results show very similar patterns. In these three studies, random failures accounted for 77–92 percent of the total failure population, and age-related failures for the remaining 8–23 percent. The authors are not aware of any other studies of a similar nature, so one can only conjecture that the age–reliability characteristics of your plant would show similar trends (which we believe is very likely to be the case).

The significance of these results, and their potential importance to the maintenance engineer, cannot be stated too strongly. Let's examine these more closely, assuming for the moment that these curves may be characteristic of your plant or system.

1. Only a very small fraction of the components (3–4 percent) actually replicated the traditional bathtub curve concept (curve A).
2. More significantly, only 4–20 percent of the components experienced a distinct aging region during the useful life of the aircraft fleets (curves A and B). If we are generous in our interpretation, and allow that curve C also is an aging pattern, this still means that only 8–23 percent of the components experienced an aging characteristic!
3. Conversely, 77–92 percent of the components never saw any aging or wearout mechanism developing over the useful life of the airplanes (curves D, E, and F). Thus, while common perceptions tend toward the belief that 9 out of 10 components have "bathtub" behavior, the analysis indicated that this trend was completely reversed when the facts were known.
4. Notice that many components, however, did experience the infant mortality phenomenon (curves A and F).

What does all of this mean? Quite a bit! First, recall that a constant failure rate region (curves A, B, D, E, and F all have this region) means that the equipment failures in this region are random in nature—that is, the state of the art is not developed to the point where we can predict what failure mechanisms may be involved, nor do we know precisely when they will occur. We only know that, on average in a large population, the instantaneous failure rate (or the mean time between failure) is a constant value. Of course, we hope that this constant failure rate value is very small, and we thus have a very reliable set of components in our system. But, for the maintenance engineer, these constant failure rate regions mean that overhaul actions will essentially (short of luck) do very little, if anything, to restore the equipment to a like-new condition. In this constant value region, overhaul is usually a waste of money because we really do not know what to restore, nor do we really know the proper time to initiate an overhaul. (In the constant failure rate region, any time you might select is essentially the wrong time!) Second, and worse yet, is that these overhaul actions may actually be harmful because, in our haste to restore equipment to new, pristine condition, we may have inadvertently pushed it back into the infant mortality region of the curve due to human error during the intrusive actions (see Figure 2.2 in Sec. 2.5). In this specific study, for example, overhaul actions on the components in curves D, E, and F would be susceptible to this counterproductive situation. A third point relates to the periodicity that should be specified for an overhaul task when such an action is considered to be the correct step to take. For example, if a component is either a curve A or B type, we want to assure that the overhaul action is not taken too soon—or again, we may be wasting our resources. Often, we do not know what the correct interval should be, or even if an overhaul PM task is the right thing to do. Why? Because we do not have sufficient data to tie down the age–reliability patterns for our equipment. In these instances, we may wish to initiate an Age Exploration program, and more on this topic will be covered in Sec. 5.9.

In summary, we should be very careful about selecting overhaul PM tasks because our equipment may not have an age–reliability pattern that justifies such tasks. In addition, due to human errors, overhauls are likely to cause more problems than they prevent if aging regions are not present. When data is absent to guide us on this very fundamental and important issue, we should initiate an Age Exploration program and/or the collection of data for statistical analyses that will permit us to make the right decisions. We should also defer, where possible, to the non-intrusive condition-directed tasks until we have more definitive results from the age exploration process. It is indeed a curious (and unfortunate) fact that, in today's world of modern technology, one of the least understood phenomena about our marvelous machines is how and why they fail!

4.3 THE BIRTH OF RCM

RCM epitomizes the old adage that "necessity is the mother of invention."

In the late 1960s, we found ourselves on the threshold of the jumbo jet aircraft era. The 747 was no longer a dream; the reality was taking shape as hardware at the Boeing factory in Seattle. The licensing of an aircraft type (called Type Certification by the FAA) requires, among its many elements, that the FAA-approved preventive maintenance program be specified for initial use by all owners/operators of the aircraft. No aircraft can be sold without this Type Certification by the FAA. The recognized size of the 747 (three times as many passengers as the 707 or DC-8), its new engines (the large, high bypass ratio fan jet), and its many technology advances in structures, avionics, and the like, all led the FAA to initially take the position that preventive maintenance on the 747 would be very extensive (like 3× the preventive maintenance on the 707 and DC-8 aircraft). This direction was so extensive, in fact, that the airlines could not likely operate this airplane in a profitable fashion.

This development led the commercial aircraft industry to essentially undertake a complete reevaluation of preventive maintenance strategy. This effort was led by United Airlines who, throughout the 1960s, had spearheaded a complete review of why maintenance was done and how it should best be accomplished. Names like Bill Mentzer, Tom Matteson, Stan Nowland, and Howard Heap, all of United Airlines, stand out as the pioneers of this effort (Refs. 16, 18, 19). What resulted from this effort was not only the thinking derived from the curves in Figure 4.1, but also a whole new approach that employed a decision-tree process for ranking PM tasks that were necessary to preserve critical aircraft functions during flight. This new technique for structuring PM programs was defined in MSG-1 (Maintenance Steering Group-1) for the 747, and was subsequently approved by the FAA. The MSG-1 was able to sort out the wheat from the chaff in a very rational and logical manner. When this was done, it was clear that the economics of preventive maintenance on a 747-sized aircraft were quite viable—and the 747 became a reality.

The MSG-1 was so successful that its principles were applied, in MSG-2, to the Type Certification of the DC-10 and L-1011. In recent times, MSG-3 has developed the PM program for the 757, 767, and 777. Versions of the MSG format have likewise guided the PM programs for the Concorde, Airbus, 737 series, and various retrofits to aircraft such as the 727-200, DC-8 stretch, and DC-9 series.

In 1972, these ideas were first applied by United Airlines under Department of Defense (DOD) contract to the Navy P-3 and S-3 aircraft and, in 1974, to the Air Force F-4J. In 1975, DOD directed that the MSG concept be labeled "Reliability-Centered Maintenance," and that it be applied to all major military systems. In 1978, United Airlines produced the initial RCM "bible" (Ref. 16) under DOD contract. More recent discussions on RCM applications in commercial aviation are found in Refs. 20 and 21.

Since then, all military services have employed RCM on their major weapons systems. RCM specifications have been developed (e.g., Ref. 22), the

Air Force Institute of Technology (AFIT) offers a course in RCM, and the Navy has published an RCM handbook (Ref. 23).

In 1983, the Electric Power Research Institute (EPRI) initiated RCM pilot studies on nuclear power plants (Refs. 24–26). Since these early pilot studies, several full-scale RCM applications have been initiated in commercial nuclear and fossil power plants (e.g., Refs. 27–31). From there, large segments of U.S. industry began to explore RCM as a basis for their maintenance improvement programs. Today (2003), many major U.S. corporations are in some stage of RCM usage, and some have revamped large portions of their PM efforts via implementation of RCM methodology.

4.4 WHAT IS RCM?

In Sec. 2.5, we briefly examined some of the more prominent practices and myths that currently constitute the basis for PM program development. We summarized by saying that these practices and myths are driven, in large measure, by the over-riding consideration and principle of "what can be done?". Until recently, this has resulted in little, if any, conscious and deliberate consideration as to why we take certain PM actions and what, if any, priority should be assigned to the expenditure of PM resources. We could further say that the overriding motivation can be simply characterized as "Preserve Equipment." Almost without fail, our current maintenance planning process starts directly with the equipment, and its sole purpose is to specify actions required to "keep it running," irrespective of the purpose or role that the equipment serves. By way of simple illustration, consider the situation where two identical air-operated valves in a nuclear power plant serve the following purposes: one regulates water flow to the main heat exchangers that provide the proper balance of steam flow to a turbine-generator set, while the other regulates service water flow to the plant facilities (e.g. cafeteria, lavatories, shops, etc.). With a "preserve equipment" mind set, both valves receive exactly the same PM actions in all likelihood. Does that make sense? Could it be that the service water valve should be run-to-failure (i.e., restored only when failure occurs)? If so, how could such decisions be made? It was questions typically like these which led people to think about a better way to use maintenance resources—and ultimately to develop the RCM process.

You should recognize from the outset that RCM is not just another cleverly packaged way to do the same old thing again. Rather, it is very different in some fundamental aspects from what today is the norm among maintenance practitioners, and requires that some very basic changes in our mindset must occur. As you will see in a moment, however, the basic RCM concept is really quite simple, and might be viewed as organized common sense.

So just what is RCM? There are four features that define and characterize RCM, and set it apart from any other maintenance planning process in use today. We will use a hypothetical scenario to develop and understand these four features.

4.4.1 Feature 1

Picture a typical business conference room which, we will hypothesize, represents the location of a system in our process plant. As we stand outside the walls (i.e., boundary) of the room, we observe that a 24 in. dia. pipe is moving water at ambient pressure and temperature into the room (i.e., system). At the other end, we find a 24 in. dia. pipe exiting the room, but now the water has been elevated in pressure and temperature to some levels that are required elsewhere in our process plant. Notice that at this point, we (theoretically) have no idea what is inside the room. But whatever this may be, it has made the room capable of elevating water pressure and temperature. We call this capability the function of the room (i.e., system), and we are able to accurately define this function without any knowledge of the room contents (i.e., equipment). In order for our plant to produce its end product, we must assure that this system continues to perform its job. That is, we must *"preserve system function"*—and this is the first and most important feature of RCM. At first glance, this is a difficult concept to accept because it is contrary to our ingrained mindset that PM is performed to preserve equipment operation. By first addressing system function, we are saying that we want to know what the expected output is supposed to be, and that preserving that output (function) is our primary task at hand. This first feature enables us to systematically decide in later stages of the process just what equipments relate to what functions, and not to assume a priori that "every item of equipment is equally important," a tendency that seems to pervade the current PM planning approach.

Let's look at a couple of simple examples to illustrate the inherent value associated with the "preserve system function" concept. First, compare two separate fluid transfer trains in a process plant where each train has redundant legs. Train A has 100 percent capacity pumps in each leg, and train B has 50 percent capacity pumps in each leg. As the plant manager, I tell you, the maintenance director, that your budget will allow PM tasks on either train A pumps, or train B pumps, but not both. What do you do? Clearly, if you don't think function, you are in a dilemma, since your background says that your job is to keep all four pumps up and running. But if you do think function, it is clear that you must devote the defined resources to the train B pumps since loss of a single pump reduces capacity by 50%. Conversely, a loss of one pump in train A does not reduce capacity at all, and also in all likelihood allows a sizeable grace period to bring the failed pump back to operation. As a second example, let's examine more closely just what function is really performed by a pump. The standard answer is to preserve pressure or maintain flow rate—a correct answer. But there is another, more subtle, function to preserve fluid boundary integrity (a passive function). In some cases, allocation of limited resources to PM tasks for the passive function could be more important than keeping the pump running (e.g., when the fluid is toxic or radioactive). Again, if you don't think function, you may miss drawing the proper attention to the passive boundary integrity function. In actual practice, virtually all passive functions are related to some structural aspect of the system and its equipment.

4.4.2 Feature 2

Since the primary objective is to preserve system function, then loss of function or functional failure is the next item of consideration. Functional failures come in many sizes and shapes, and are not always a simple "we have it or we don't" situation. We must always carefully examine the many in-between states that could exist, because certain of these states may ultimately be very important. The loss of fluid boundary integrity is one example of a functional failure that can illustrate this point. A system loss of fluid can be (1) a very minor leak that may be qualitatively defined as a drip; (2) a fluid loss that can be defined as a design basis leak—that is, any loss beyond a certain GPM value will produce a negative effect on system function (but not necessarily a total loss), and (3) a total loss of boundary integrity, which can be defined as a catastrophic loss of fluid and loss of function. In this example, then, a single function (preserve fluid boundary integrity) led to three distinct functional failures.

Thus, in our hypothetical system described above, we could measure flow rate at the exit of the pipe, and determine if any one of the above three functional failures might be present. Or we may be able to discern a small leak by simply observing fluid loss across the boundary (e.g., water on the floor). If yes, the paramount question, then, would be to ascertain just what has happened inside the room to produce the functional failure. To answer this question, we now open the door and step into the room (system). There before us are all of the components (equipment) that are working together in some harmonious manner to produce the function that was observed when we were standing outside the walls (boundary) of the room. Our job now is to meticulously examine each component in order to delineate just how it might fail such that the functional failure(s) could occur. Thus, the key point in Feature 2 is that we make the transition to the hardware components by *"identifying specific failure modes that could potentially produce the unwanted functional failures."* By way of illustration, a flow control valve (component) that is jammed shut (failure mode) could produce the functional failure "fails to initiate system startup."

4.4.3 Feature 3

In the RCM process, where our primary objective is to preserve system function, we have the opportunity to decide, in a very systematic way, just what order or priority we wish to assign in allocating budgets and resources. In other words, "all functions are not created equal," and therefore all functional failures and their related components and failure modes are not created equal. Thus we want to *"prioritize the importance of the failure modes."* This is done by passing each failure mode through a simple, three-tier decision tree which will place each failure mode in one of four categories that can then be used to develop a priority assignment rationale. (This will be discussed in detail in Sec. 5.7.)

4.4.4. Feature 4

Notice that, up to this point, we have not yet dealt directly with the issue of any preventive maintenance actions. What we have been doing is formulating a very systematic roadmap that tells us the where (component), what (failure mode), and priority with which we should now proceed in order to establish specific PM tasks—all of this being driven by the fundamental premise to "preserve function." We thus address each failure mode, in its prioritized order, to *identify candidate PM actions* that could be considered. And here, RCM again has one last unique feature that must be satisfied. Each potential PM task must be *judged as being "applicable and effective."* Applicable means that if the task is performed, irrespective of cost, it will in fact accomplish one of the three reasons for doing PM (i.e., prevent or mitigate failure, detect onset of a failure, or discover a hidden failure). Effective means that we are willing to spend the resources to do it. Generally, if more than one candidate task is judged to be applicable, we would opt to select the least expensive (i.e., most effective) task. Recall that in Sec. 2.3, when describing a run-to-failure task category, we indicated three reasons for such a selection. We can now be more precise, and state that failure of a task to pass either the applicability or effectiveness test results in two of the run-to-failure decisions. The third would be associated with a low-priority ranking and a decision not to spend any PM resources on such insignificant failure modes.

4.4.5 The Four Features—A Summary

In summary, then, the RCM methodology is completely described in four unique features:

1. Preserve functions.
2. Identify failure modes that can defeat the functions.
3. Prioritize function need (via failure modes).
4. Select applicable and effective PM tasks for the high priority failure modes.

These four features or principles are actually implemented in a systematic, stepwise process that is described in detail in Chapter 5.

The above four features totally describe the RCM concept—nothing more and nothing less. For any maintenance analysis process to be labeled as RCM, it must contain all four features. The authors have occasionally encountered maintenance programs that are purported to be RCM programs but lack one or more of the four features. And usually these programs are also less than satisfactory, and tend to give RCM an unfair reputation. So we caution you to avoid the shortcuts if you truly wish to have an RCM-based PM program.

4.5 Some Cost–Benefit Considerations

As noted earlier, the primary driving force behind the invention of RCM was the need to develop a PM strategy that could adequately address system availability and safety without creating a totally impractical cost requirement. This has clearly been successfully achieved by commercial aircraft; however, quantitative commercial airline data on this cost picture in the public arena is rather scarce. Figure 4.3 (Ref. 32) presents maintenance cost per flight-hour in the first 10 years of RCM use. This figure was originally displayed in the mid-1970s to illustrate the impact of the OPEC oil embargo crisis on the escalating cost of airline operations. But it also serves well in illustrating the trend in maintenance costs. What we see in Figure 4.3 is a maintenance cost that is essentially constant from the late 1960s to the early 1980s. This is precisely the period during which the 747, DC-10, and L-1011 were introduced into revenue service on a large scale. These jumbo jets introduced not only a large increase in passenger capability per aircraft (about three times over that of the 707 and DC-8), but also a higher daily usage rate and the deployment of several advanced technologies into everyday use. In spite of all of these factors, any one of which would normally tend to drive maintenance cost up, we see a fairly constant maintenance cost per flight-hour historically occurring. RCM was the overriding reason for this.

Figure 4.4 (Refs. 20, 32) presents another way to view the impact of RCM in the commercial aircraft world. Note that the PM definitions used in Figure 4.4 correspond as follows to the PM task definitions given in Sec. 2.3:

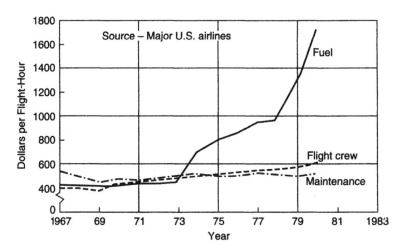

Figure 4.3 Costs per flight-hour (1982 constant dollars).

| Maintenance | Component distribution | | |
process	1964	1969	1987 (est.)
Hard-time* units	58%	31%	9%
On-condition† units	40%	37%	40%
Condition-monitored‡ units	2%	32%	51%

   * *Hard-time*—Process under which an item must be removed from service at or before a previously specified time.
   † *On-condition*—Process having repetitive inspections or tests to determine the condition of units with regard to continued serviceability (corrective action is taken when required by item condition).
   ‡ *Condition-monitored*—Process under which data on the whole population of specified items in service is analyzed to indicate whether some allocation of technical resources is required. Not a preventive maintenance process, CM *allows failures to occur,* and relies upon analysis of operating experience information to indicate the need for corrective action.
   NOTE: Definitions from *World Airlines Technical Operations Glossary*—March 1981.

Figure 4.4 Commercial aircraft—component maintenance policy.

Hard-time	Time-directed
On-condition	Condition-directed
Condition-monitored*	Run-to-failure

Two significant points can be observed with this data. First, the pre- (1964) and post- (1969, 1987) RCM periods reveal the dramatic shift that occurred in the reduction of costly component overhauls (i.e., hard-time tasks), mainly in favor of run-to-failure (i.e., condition-monitored) tasks. Much of this shift, of course, was made possible by a design philosophy that included double and triple redundancy in the flight-critical functions. The RCM process was employed to take advantage of these design features in structuring where PM was critical and where the run-to-failure decision was appropriate. Also, notice that throughout the time period represented here the condition-directed (i.e., on-condition) task structure remained fairly constant. The commercial aircraft industry was one of the early users of performance and diagnostic monitoring as a PM tool, and has continued to apply it successfully throughout the generation of the newer jet aircraft.

The results shown in Figures 4.3 and 4.4 slowly led to a growing interest in RCM applications in other commercial areas. The first comprehensive commercial look at RCM in the U.S. was triggered by the Electric Power Research Institute (EPRI) in its study of commercial aviation practices with application to nuclear power

---

*While the airlines glossary definitions treat actions resulting from run-to-failure as corrective maintenance, the authors hold to the definition that RTF is a preventive maintenance option that is deliberately scheduled only when failure has occurred, and thus repair/replace actions are PM actions.

plants (Ref. 32). This milestone study led to pilot RCM projects at Turkey Point (FPL) and McGuire (Duke) nuclear plants (Refs. 24, 25) in the early 1980s. This opened the floodgates for additional RCM projects throughout the electric utility industry, and from there to the spread of RCM into several commercial areas such as petrochemical, process, and manufacturing plants throughout the U.S. In a parallel vein, many foreign companies were also moving into the use of RCM through the efforts of John Moubray and his RCM2 publication (Ref. 33), which utilizes the same four features described above. All of these efforts were primarily focused on reduction of both O&M costs and plant downtime.

Cost payback from an upfront investment in RCM has been profoundly successful. In the early 1990s, EPRI conducted a survey of seven U.S. electric utilities and found that an average payback period of 6.6 years had been demonstrated in the early RCM projects, with payback expected to quickly reach two years (Ref. 34). The authors' experience is that the cost payback is consistently less than one year, and in many cases the payback is almost instantaneous due to the peripheral benefits realized from the RCM IOI analysis process in areas such as design and operational modifications that were made on the basis of RCM information (see Sec. 5.10).

Further discussion and illustrations of RCM program costs versus return-on-investment (ROI) are presented in Sec. 9.1—The Financial Factor. Seven case history studies, including "Benefits realized," are presented in Sec. 12.2.

5

RCM METHODOLOGY—THE
SYSTEMS ANALYSIS PROCESS

This chapter will provide a comprehensive description of the systems analysis process that is used to implement the four basic features which define and characterize RCM (see Sec. 4.4). This process will be discussed in terms of seven steps that have been developed from experience as a most convenient way to systematically delineate the required information:

Step 1: System selection and information collection.
Step 2: System boundary definition.
Step 3: System description and functional block diagram.
Step 4: System functions and functional failures—Preserve functions.
Step 5: Failure mode and effects analysis (FMEA)—Identify failure modes that can defeat the functions.
Step 6: Logic (decision) tree analysis (LTA)—Prioritize function need via the failure modes.
Step 7: Task selection—Select only applicable and effective PM tasks.

Satisfactory completion of these seven steps will provide a baseline definition of the preferred PM tasks on each system with a well-documented record of exactly how those tasks were selected and why they are considered to be the best selections among competing alternatives. Notice that Steps 4, 5, 6, and 7 correspond to the four basic features that characterize RCM. Two additional steps are required in order to complete a successful RCM Program:

Step 8: Task packaging—which will carry the recommended RCM tasks to the floor (Chapter 8 is devoted to this effort).

71

Step 9: Living RCM program—comprising the actions necessary to sustain over time the beneficial results of Steps 1–8 (Chapter 10 describes this effort).

A complete, albeit simple, systems analysis process using Steps 1–7 above is illustrated in Chapter 6, using a swimming pool as the example. Seven case studies drawn from actual use of this systems analysis process are described in Chapter 12.

The systems analysis process described in this chapter is often referred to as the Classical RCM process, and is so named because this is the process that most closely follows the RCM methodology that was first defined and successfully employed in MSG-1 for the 747 airplane.

5.1 SOME PRELIMINARY REMARKS

When an analyst embarks upon the process described in this chapter, it is helpful to keep a few key points in mind relative to the application of the RCM systems analysis process.

1. Traditional methods for determining PM tasks start with the issue of preserving equipment operability, and such methods tend to focus the entire task selection process on what can be done to the equipment. As a rule, why it should be done is never clearly addressed (or documented, if such consideration was, in fact, ever investigated). RCM is a major departure from this traditional practice! Its basic premise is "preserve function"—not "preserve equipment." This approach forces the analyst to systematically understand (and document) the system functions that must be preserved without any specific regard initially as to the equipment that may be involved. It then requires the analyst to think carefully about how functions are lost—in functional failure terms, not equipment failure terms. The purpose of this approach is to develop a credible rationale for why one might eventually desire to perform an appropriate PM task rather than just arbitrarily deciding to do something because "it sounds right." (The "preserve function" approach is initially developed in Steps 3 and 4.)

2. However, this is not to imply that traditional experience and sound engineering judgment about equipment malfunctions is unimportant to the RCM process. On the contrary, the use of operations and maintenance personnel experience, as well as historical data from plant-specific and generic data files that are properly screened for your application, is an invaluable input to assuring that all important failure modes are eventually captured and considered in the FMEA (Step 5).

3. The direct involvement of plant operations and maintenance personnel in the RCM systems analysis process is extremely important from another point of view also—namely, as a "buy-in" to the process. This embodies

and promotes a feeling of belonging, and satisfies a very real necessity for them to share in the formulation of the PM tasks that they will eventually be asked to implement. Experience with many RCM programs has shown that success is rarely achieved if the buy-in factor has been neglected.

4. Recall from Sec. 2.3 that there are four categories from which to choose a candidate PM task: (1) time-directed (TD), (2) condition-directed (CD), (3) failure-finding (FF), and (4) run-to-failure (RTF). As a rule, there is virtually no difficulty with people accepting the definition and use of TD and CD tasks in a PM program. Use of the FF task as a formal inclusion in the PM program is new to most people, but is generally accepted as a valid PM task in a short period of time. But the notion of a deliberate decision to run-to-failure is totally foreign to the more traditional elements of preventive maintenance, and frequently becomes a very difficult concept to sell. Thus, some care and sensitivity to the use of RTF tasks is necessary, and may entail some special education efforts to ensure that the operations and maintenance personnel understand why RTF tasks are, in fact, the best selection. The specific reasons behind RTF are developed in Steps 5, 6, and 7.

5. It is not uncommon for people receiving their first exposure to Classical RCM to comment that "there sure is a lot of paperwork involved here." And in the framework, say, of a plant maintenance director and his or her staff, there is some truth in the comment. Thus, it becomes important to emphasize certain crucial points in order to help people to understand why the paperwork is there, and how it benefits them in the long run. These points should include the following:

- RCM wants to ensure that you can answer, both today and in the future, the "why" behind every task that will use your limited resources (i.e., preserve the most important functions). It can be especially important to know why you may have elected not to follow an OEM recommendation since regulators and insurers often tend to hold them in high regard.
- RCM wants to ensure that your task selections derive from a comprehensive knowledge of equipment failure modes because it is at that level of detail where failure prevention, detection, or discovery must occur. If your task selection process, whatever it may be, does not do this, then there is no assurance that the task really does anything particularly useful.
- RCM wants to ensure that the most effective (least costly) task is chosen for implementation. Historically, this has not been done and, consequently, most PM programs fall far short of realizing the best return for the resources spent.
- Under certain conditions more specifically defined in Step 1 (Sec. 5.2), alternative methods of analysis are described which reduce the manhours required to optimize a PM program (see Chapter 7 for details).

In order to realize these benefits, it does take some effort and documentation. But, once a system has been through the RCM process, it produces a baseline definition of the PM program for that system which needs only periodic update to account for new information and system changes (see Chapter 10). Thus, the systems analysis process is, in fact, a one-shot process that thoroughly documents where you are and why—a point of increasing concern in the current economic and regulatory climate. Further, as the RCM process has evolved and matured, much of the mechanics of the analysis has been computerized, and this has introduced efficiencies as well as eliminated the need for hard copy reports where such was desired. The subject of software support for the RCM process is discussed in Chapter 11.

6. It should be noted that the RCM methodology focuses only on what task should be done and why (i.e., task definition). All tasks must likewise establish *when* the task should be done (i.e., task frequency or periodicity), but these intervals are derived from separate analyses that must consider and utilize combinations of company and industry experience to establish initial task frequencies. More sophisticated statistical tools may be employed when data is available to pursue this avenue; also, controlled measurement techniques known as Age Exploration can be used. More will be said on this in Sec. 5.9.

7. If possible, the systems analysis process should involve a team of two or three analysts and one of them should be from operations. This will encourage not only cross-talk about what information should be included in each of the seven steps, but also a healthy level of challenge, questioning, and probing in that regard. Further, in such team arrangements, it is beneficial to include one team member who is not totally conversant in the system under investigation and one who is experienced. This will also help to develop the cross-talk and challenge process to the betterment of the end product. Specific thoughts about the formulation of RCM teams are given in Sec. 9.2.

8. Lastly, it is important that proper consideration be given to scheduling the activities of the RCM team in order to minimize any impact on the normal responsibilities of the team members. This is further discussed in Sec. 9.3.

Experience has shown that the analyst will be placed in a more proper state of mind to proceed with an RCM systems analysis if some initial appreciation of the above points has been acquired and accepted.

5.2 Step 1—System Selection and Information Collection

When a decision has been made to perform an RCM program at your plant or facility, two immediate questions arise:

1. At what level of assembly (component, system, plant) should the analysis process be conducted?

2. Should the entire plant/facility receive the process—and, if not, how are selections made?

5.2.1 Level of Assembly

For our discussion here, we can think of the following definitions to describe levels of assembly:

- Part (or piece part): the lowest level to which equipment can be disassembled without damage or destruction to the item involved. Items such as microprocessor chips, gaskets, ball bearings, gears, and resistors are examples of parts. Notice size is not a criterion in this regard.
- Component (or black box): a grouping or collection of piece parts into some identifiable package that will perform at least one significant function as a stand-alone item. Often, modules, circuit boards, and subassemblies are defined as intermediate buildup levels between part and component. Pumps, valves, power supplies, turbines, and electric motors are typical examples of components. Again, size is not a criterion.
- System: a logical grouping of components that will perform a series of key functions that are required of a plant or facility. As a rule, plants are composed of several major systems such as feedwater, condensate, steam supply, air supply, water treatment, fuel, and fire protection.
- Plant (or facility): logical grouping of systems that function together to provide an output (e.g., electricity) or product (e.g., gasoline) by processing and manipulating various input raw materials and feedstock (e.g., water, crude oil, natural gas, iron ore).

When PM planning is approached from the function point of view, experience has rather clearly shown that the most efficient and meaningful function list for RCM analysis is derived at the system level. In most plants or facilities, the systems have usually been identified, since they are also used as logical building blocks in the design process, and plant schematics and piping and instrumentation diagrams (P&IDs) thus define these systems rather precisely. These system definitions typically serve well as a starting point for the RCM process.

The authors have supported some clients with their RCM program where the plant has not been designed as an aggregation of discretely defined systems. Some manufacturing and process plants (for example, production lines in the aircraft and paper industry) represent such a situation. In these cases, we have found that it is relatively easy to synthesize the partition of the plant into equivalent systems in order to simplify the systems selection process. We have also found that, in special circumstances, a single large, unique, or specialized asset may serve as a system—for example, a diesel-generator set, a large air compressor, or a complex machine tool.

A reasonable way to explain and justify the use of systems in the RCM process is to consider alternatives—that is, why not components on a one-by-one basis? Or, at the other extreme, why not the entire plant in a single analysis process? First, at the component level, it becomes difficult, sometimes impossible:

> —to define the significance of functions and functional failures. For example, a valve can fail to open or fail to close on demand, and defeat some flow function that it controls; but unless the analyst looks more broadly at the system functions that are affected, we may not truly know the significance of the component function. We will also find later that a single component often supports several functions, and this becomes clear to the analyst only when viewing the entire system, not just one component; and
> —to perform meaningful priority rankings between failure modes that are competing for limited PM resources. In a component we may have only two, or at most six to eight, failure modes to compare, whereas a system typically has hundreds, and comparisons make more sense.

At the other end of the spectrum, quite simply, the entire plant in one bite will literally choke the analysis process, and create an analysis nightmare in attempting to follow too many functions at once. Even combining two systems in one analysis (in a trial case, condensate and feedwater in a power generation plant) proved to be extremely cumbersome and difficult to track, and was abandoned in favor of two separate systems analyses before Step 4 was completed. Generally, it can be stated that multiple systems analysis packages tend to exceed the cumulative time required to perform separate systems analyses because of the confusion created with the multiple system approach.

To summarize, then, the recommended approach is to conduct the RCM analysis process at the systems level—hence the term systems analysis process.

5.2.2 System Selection

Having established that the system is the best practical level of assembly at which to conduct the Classical RCM analysis process, we can now focus on which systems to address and in what order. Obviously, one decision could be to treat all plant/facility systems. However, we have consistently found that such a course of action may not be cost-effective from a maintenance viewpoint in that some systems have neither a history of frequent failures, excessive maintenance costs, nor contributions to forced outages that might warrant a special investigation to "make it better." Given that such may be the case in your plant or facility, what procedure might be employed to select those systems with the highest potential for benefit from the Classical RCM systems analysis process?

The most direct and credible way to answer this question is to invoke the 80/20 rule. This concept was first mentioned in Sec. 1.2, Item #11, and was described as a rule which states that 80% of an observed effect tends to reside in 20% of the available source. In our case, the observed effect of interest revolves around the high cost of maintenance and/or a large amount of plant downtime while the available source is one or more of the plant systems (i.e., the bad-actor systems). A more comprehensive, and interesting, treatise on the 80/20 rule can be found in Ref. 7. To use the 80/20 rule as the basis for system selection, we need to assemble data that will represent maintenance costs or downtime, on a system-by-system basis, and plot this information in a Pareto diagram (a bar chart plotted in a descending order of value). The authors have consistently used one or more of the following parameters to construct these Pareto diagrams:

1. Cost of corrective maintenance actions over a recent two-year period.
2. Number of corrective maintenance actions over a recent two-year period.
3. Number of hours attributed to plant outages over a recent two-year period.

This data, segregated by system, is readily available in even the most primitive plant data systems. A typical Pareto diagram using the equivalent of outage hours converted to an effective forced outage rate (EFOR) is shown in Figure 5.1 for a fossil power generation plant at Florida Power & Light. This diagram can be used to easily determine where approximately 80% of the EFOR resides, thus defining the first three systems as the "bad actors." That is, the greatest ROI can be realized by concentrating on just three of the 11 systems shown on the diagram.

Another Pareto diagram based on maintenance costs for a major test facility at the USAF Arnold Engineering Development Center is shown in Figure 5.2. Both Figures 5.1 and 5.2 illustrate the power of the 80/20 rule in clearly identifying just where the best opportunity for significant ROI resides.

In selected cases, we have used all three of the above-listed parameters to construct an 80/20 solution, and found that all three Pareto diagrams provided essentially the same list of system selections with some slight variation only in the order in which they appeared. This may be important for you because #2 above, the number of corrective maintenance actions, may be the easiest data to retrieve in order to construct your Pareto diagram.

Two caveats in this procedure deserve some attention as observed from actual practice. First, it may be possible that one of the selected systems is a large contributor to maintenance actions or downtime because of a single problem that is related to the need for a design modification. In one such instance, the design modification was introduced in the 6-month time period immediately following the two-year period that had been selected for the Pareto diagram data. And in this later 6-month period, the problem had been corrected and the system was now considered to be among the "best" in the plant. Quite obviously, pursuit of

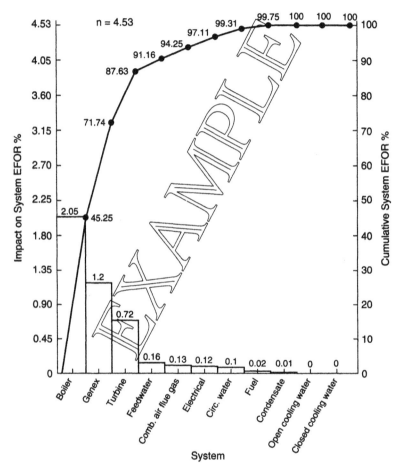

FPL Generating System EFOR % Impact
By Designated Plant System
1984 – 1987

Figure 5.1 Typical EFOR Pareto diagram (courtesy of Florida Power & Light).

this system to optimize its PM schedule would have been a zero ROI exercise. A second caveat concerns a system that is predominantly composed of digital electronic hardware. The issue here is not so much that it really is causing serious downtime in your plant, but rather that this type of hardware is, for the most part, not really receptive to any type of PM except RTF. For example, do you do PM on your TV, DVD, washer/dryer controls? You get the picture. So don't be led into

Factored Cost Failure Data

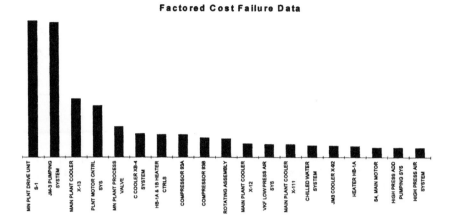

Figure 5.2 Typical maintenance cost Pareto diagram (courtesy of USAF/AEDC).

an RCM project to "solve the bad actor" digital control system in your plant. What you need is either a redesign effort or a good failure history to define your spares requirements so that the failed module or circuit board can be quickly replaced when necessary. You don't need RCM to figure that out!

In summary, the Classical RCM process should be done at the system level. In selecting which systems to use, the 80/20 rule has proven to be a consistent and credible model for assuring that the best ROI has been chosen. For the more well-behaved systems (i.e., the 20/80 systems), alternative analysis methods are described in Chapter 7.

5.2.3 Information Collection

Considerable time and effort can be saved (and perhaps continuing frustration avoided) by researching and collecting, at the outset, some necessary system documents and information that will be needed in subsequent steps. A list of the documents and information typically required for each system by the RCM analysts is as follows:

1. System piping and instrumentation diagram (P&ID). See example in Figure 5.3 (from Ref. 24).
2. System schematic and/or block diagram. Frequently, these are developed from the P&ID to help in the visual display of how the system works, and usually are less cluttered than the P&ID, thus facilitating a good understanding of the main equipment and function features of the system. See example in Figure 5.4.

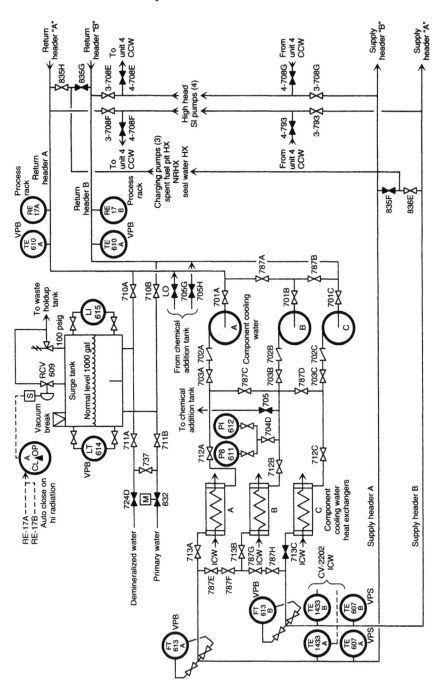

Figure 5.3 Component cooling water P&ID.

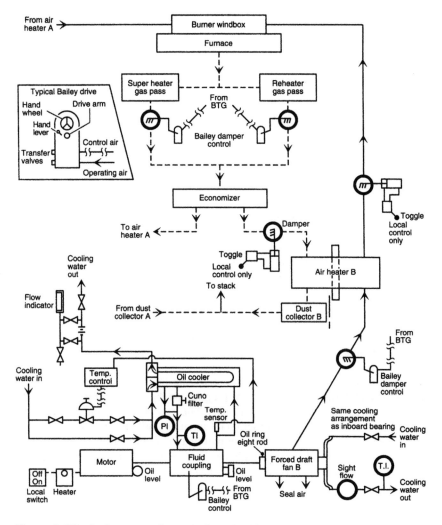

Figure 5.4 Typical system schematic for a fossil power plant (courtesy of Florida Power & Light).

3. Individual vendor manuals for the equipments in the system, which will contain potentially valuable information on the design and operation of the equipments for use in Step 5 (FMEA).
4. Equipment history files, which will list the actual failures and corrective maintenance actions that have occurred in your facility for documentation in Step 3-5 and for use in Step 5.

5. System operation manuals, which will provide valuable details on how the system is intended to function, how it relates to other systems, and what operational limits and ground rules are employed. These items are of direct use in Step 4 ("System functions and functional failures").
6. System design specification and description data, which will generally support and augment all of the preceding and, most importantly, will help to identify information needed in Step 3-1 ("System functional description") and Step 4.

There may be other sources of information that are unique to your plant or organization structure that would be helpful to accumulate. Also, industrywide data (such as equipment failure histories) are useful, when available, to augment your own experience. As you can see, all of the preceding items are aimed at assuring that the analyst has sufficient detail to thoroughly understand what is in the system, how it works, and what has been the historical equipment experience.

Suffice it to say that you may not always have all of these items. For example, there may be no P&ID available, and you may have to "create" one via a system walkdown and visual reconstruction of the as-built configuration. Or you may have to conduct interviews to ferret out the equipment history. In older plants and facilities, this is frequently necessary. Even when the documentation is complete, system walkdowns and staff interviews are good ideas, and more will be said on these two points in later discussions.

One caveat: an item missing from the preceding list is the collection of documentation that defines the existing PM program on the system. This is eventually needed in Step 7.3 ("Task Comparisons"), but it is not recommended that the analyst acquire the current PM program information until Step 7.3 in order to preclude any prior knowledge on his or her part that might influence or bias the decisions on what the RCM results should be. There will be plenty of opportunity in the later stages of Step 7 to collect this data and specifically compare it to the RCM results.

5.3 STEP 2—SYSTEM BOUNDARY DEFINITION

The number of separately identifiable systems in a plant or facility can vary widely, depending upon plant or facility complexity, financial accounting practices, regulatory constraints, and other factors that may be unique to a given industry or organization. For example, in the power generation industry, an 800 MWe fossil (oil, gas, or coal-fired) plant typically will have about 15 to 30 separate systems, whereas an 800 MWe nuclear plant may well have in excess of 100 separate systems. Some gross system definitions and boundaries usually have been established in the normal course of the plant or facility design, and these system definitions have already been used in Step 1 as the basis for system selection.

These same definitions serve quite well to initially define the precise boundaries that must be identified for the RCM analysis process.

Why is precise system boundary definition so important in the RCM analysis process? There are two reasons:

1. There must be a precise knowledge of what has or has not been included in the system so that an accurate list of components can be identified or, conversely, will not overlap with components in an adjacent system. This is especially true when RCM analyses are to be performed on two adjacent systems which, in all likelihood, will be done at different times and may involve different analysts.
2. More importantly, the boundaries will be the determining factor in establishing what comes into the system by way of power, signals, flow, heat, etc. (what we call the IN interfaces) and what leaves the system (the OUT interfaces). As will be discussed in Steps 3 and 4, a clear definition of IN and OUT interfaces is a necessary condition to assuring accuracy in the systems analysis process, especially with respect to the identification of all system functions. This, in turn, depends upon a clear understanding of what is or is not included in the system. That is, where have the system boundaries physically been established?

In Sec. 4.4, we described the four RCM features by talking about a conference room that was hypothesized to represent a plant system. It should be obvious that without a definitive knowledge of where the conference room walls are located, we would have some difficulty in accurately saying just what is inside the room (i.e., components), and what goes in and out of the room (i.e., IN and OUT interfaces). Our experience is that this simple fact seems to escape many people, and the result is that they want to drop Step 2! Several years ago, we ran an experiment. We took equally knowledgeable RCM teams, and asked both of them to (a) define the components inside of System X, and (b) define the OUT interfaces for System X. Team #1 was allowed to perform Step 2: team #2 could not. Result: team #2 missed 20% of the component list and 25% of the OUT interface list while team #1 was 100% accurate. The message here, we believe, is that Step 2 provides necessary information, especially when dealing with an 80/20 system.

There are no hard and fast rules that precisely govern the establishment of system boundaries. Systems, by definition, usually have one or two top-level functions and a series of supporting functions that constitute a logical grouping of equipments. But considerable flexibility is allowed in defining precise boundary points to allow the analyst to group equipments in the most efficient manner for analysis purposes. Some examples of how this could occur will serve to best illustrate this latter point.

1. A heat exchanger may physically be in system A, but its fluid level sensors are the key input to the control of flow in system B. Hence, the level

sensors are placed in system B so that a complete picture of flow control in system B is possible.

2. An equipment lubrication function may reside in system A, but it ultimately services lubrication needs in several other systems. Here, it may be prudent to treat this lubrication function, in its entirety, as a completely separate system B.

3. System A may have control readouts in a plant control room that is physically far separated from system A. But the analyst may deem it best to include those control room instruments in this treatment of system A. Thus, if the control room is later analyzed as a separate system, the previously established boundary for system A would tell the analyst not to include those instruments in the control room boundary definition.

4. Other equipment items, such as circuit breakers (CB), can also be used as boundary points with either the entire CB or only one side of the CB within the system boundary.

Whatever decisions are reached on boundary definitions, they must be clearly stated and documented as a part of the analysis process. This is done with a two-step documentation using the forms shown on Figures 5.5 and 5.6.* Figure 5.5 gives a "Boundary overview," Step 2-1, and, as the name implies, it gives a quick indication of the major system components and the key boundary points. In some cases, it is prudent to include "Caveats of note" to warn others who may later use the analysis results about special inclusions or exclusions that were used. The data in Figure 5.5 is frequently useful in reviewing an RCM project with management. But Figure 5.6 gives very specific "Boundary details," Step 2-2, by listing each interface location with a neighboring system using a word description, reference drawing number (if applicable), and the name of the bounding system. If applicable, it also indicates whether each interface location is also a point where something enters or leaves the system (i.e., an IN or OUT interface). Not every boundary point will necessarily also be an IN or OUT interface. System boundaries can also be represented pictorially by using a highlight marker to color the system's lines on a piping and instrumentation diagram (P&ID), or by highlighting the boundary on a system schematic such as that shown in Figure 5.7.

As the analyst proceeds with Steps 3 and 4, it may become evident that the system boundary needs some adjustment to accommodate factors not originally envisioned. This is an acceptable practice, and in fact usually occurs as a part of an

---

*The documentation forms shown throughout Chapters 5 and 6 are those used in the "RCM WorkSaver" software (see Sec. 11.5). If this software is not used, then the RCM team should use documentation forms similar to those shown here. Prior to the creation of RCM WorkSaver, the authors used similar forms, and they were completed in handwritten format.

```
┌────────────────────────────────────────────────────────────────────────────────┐
│                         RCM - Systems Analysis                                   │
│  Step 2-1:      System Boundary Definition          Plant ID:                    │
│  Information:   Boundary Overview                    System ID  00651-020304      │
│  Plant:         VKF HPA Auxiliary Plant              Rev No:    0                 │
│  System:        JM3 Pumping System                  Date:      2/20/98           │
│  Subsystem:     C92 Compressor System                                            │
│  Analysts:      Ed Ivey, Brian Shields, Brown Limbaugh, Ronnie Skipworth, Glenn  │
│                 Hinchcliffe (facilitator)                                        │
└────────────────────────────────────────────────────────────────────────────────┘
```

Major Equipment Included:

GE 1250 Hp 6900V Induction Motor
Ingersal- Rand 3 Stage Centrifugal Compressor
Coupling
Lube Oil Pump
Pre-Lube Pump/Motor
Lube Oil Cooler
Inlet Air Filter
V921,V928, V925, Vent Valve

Primary Physical Boundaries

Start with:

Air from atmosphere entering into the filter

Terminate with:

38 PSIG air at approximately 100F on outlet side of V925

Vent excess air to atmosphere through outlet of vent line

Caveats:

Did not include any electrical supply breaker, starter, or cables in this analysis

System: JM3 Pumping System	Sunday, June 08, 2003
Subsystem: C92 Compressor System	Page 1 of 1

Step 2-1 Boundary Overview
JMS Software

Figure 5.5 Typical boundary overview on form for Step 2-1 (courtesy of USAF/AEDC).

RCM - Systems Analysis

Step 2-2:	System Boundary Definition	**Plant ID:**	
Information:	Boundary Details	**System ID**	00651-020304
Plant:	VKF HPA Auxiliary Plant	**Rev No:**	0
System:	JM3 Pumping System	**Date:**	2/20/98
Subsystem:	C92 Compressor System		
Analysts:	Ed Ivey, Brian Shields, Brown Limbaugh, Ronnie Skipworth, Glenn Hinchcliffe (facilitator)		

Type	Bounding System	Interface Location	Reference Drawing
OUT (Air)	93A/B Compressor	Down stream side of V925	20-00054.18
IN (Air)	Atmosphere	Up stream side of inlet filter	20-00054.18
OUT (Air)	Atmosphere	Outlet of ducting	20-00054.18

System:	JM3 Pumping System	Sunday, June 08, 2003
Subsystem:	C92 Compressor System	Page 1 of 1

Step 2-2 Boundary Details
JMS Software

Figure 5.6 Typical boundary details on form for Step 2-2 (courtesy of USAF/AEDC).

iteration process that will occur throughout Steps 2, 3, and 4 in order to get the most efficient results before proceeding to Step 5.

5.4 STEP 3—SYSTEM DESCRIPTION AND FUNCTIONAL BLOCK DIAGRAM

With our system selection complete, and the boundary definitions established for the first system to be analyzed, we now proceed in Step 3 to identify and document the essential details of the system that are needed to perform the remaining steps in a thorough and technically correct fashion. Five separate items of information are developed in Step 3:

- System description
- Functional block diagram
- IN/OUT interfaces
- System work breakdown structure (i.e., component list)
- Equipment history

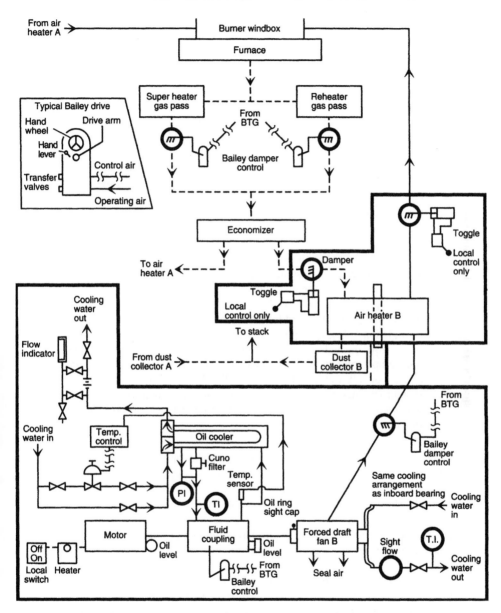

Figure 5.7 Typical system schematic indicating system boundaries (from Figure 5.4).

5.4.1 Step 3-1—System Description

By this point in the analysis process, a great deal of information has been collected and, to some degree, digested regarding what constitutes the system and how it operates. The analyst will now commit this information to the forms used in Step 3 to document the baseline definition and understanding that is used to ultimately specify PM tasks. The first item of information is the system description (Step 3-1), which is documented on the form shown in Figure 5.8. A well-documented system description will produce several tangible benefits:

1. It will help to record an accurate baseline definition of the system as it existed at the time of the analysis. Since design and operational changes in the form of modifications or upgrades can occur over time, the system must be baselined to identify where PM revisions might be required in the future (see Chapter 10—"The Living RCM Program").
2. It will assure that the analysts have, in fact, acquired a comprehensive understanding of the system. (It is rare that the analysts are "experts" in more than two or three systems.)
3. Most importantly, it will aid in the identification of critical design and operational parameters that frequently play a key role in delineating the degradation or loss of required system functions. For example, cooling-water flow through a heat exchanger may have several "allowable" inlet temperature and/or flow-rate conditions that correspond to varying degrees of "allowable" equipment cooling levels in successive states of system (or plant) capacity reduction short of a complete shutdown. Knowledge such as this becomes vital in the accurate specification of functional failures in Step 4.

The level of detail found in system descriptions varies greatly from analyst to analyst. Suffice it to say that a well-documented form (Figure 5.8) will pay significant dividends throughout the analysis process. Notice that Figure 5.8 deliberately highlights the callouts for system redundancy features (e.g., on-line standby pumps, serial "normally closed" valves, alternative modes of operation, grace periods), protection features (e.g., alarms, interlocks, and trips), and key instrumentation features (e.g., digital or analog control devices).

As an option, the analysts could elect to delay the system description until the functional block diagram (Step 3-2) has been developed to assess the possibility that subsystems may be required. This breakout into subsystems is fairly common. For example, a system divided into four subsystems may have only one or two subsystems analyzed via the Classical RCM process since these subsystems represent the majority of the data on the Pareto diagram. If this is done, the remainder of the systems analysis process is conducted at the subsystem level—i.e., when the word "system" is used, substitute "subsystem."

```
┌─────────────────────────────────────────────────────────────────────┐
│                        RCM - Systems Analysis                       │
│  Step 3-1:      System Description/Functional Block Diagram    Plant ID: │
│  Information:   System Description                             System ID: │
│  Plant:                                                       Rev No:   │
│  System:                                                      Date:     │
│  Subsystem:                                                            │
│  Analysts:                                                            │
│                                                                       │
└─────────────────────────────────────────────────────────────────────┘
```

<u>Functional Description/Key Parameters</u>

<u>Redundancy Features</u>

<u>Protection features</u>

<u>Key Control Features</u>

| System: | Tuesday, April 22, 2003 |
| Subsystem: | Page 1 of 1 |

Step 3-1 System Description
JMS Software

Figure 5.8 Form for system description, Step 3-1.

5.4.2 Step 3-2—Functional Block Diagram

The functional block diagram is a top-level representation of the major functions that the system performs and, as such, the blocks are labeled as the functional subsystems for the system. As the name denotes, this block diagram is composed solely of functions; no component or equipment titles appear in this block diagram. Typical functional subsystems might include blocks for pumping or flow, heating, cooling, mixing, cutting, lubrication, control, protection, storage, and distribution. Arrows connect the blocks to broadly represent how they interact with each other, and also to visually show the IN/OUT interfaces of the subsystems and system to complete a functional picture of what our system is supposed to do (see Step 3-3 for details). As you can probably envision at this point, the completed functional block diagram becomes a key link to Step 4, where we will formally identify and document the system functions.

The functional block diagram is recorded on the form shown in Figure 5.9, which illustrates a typical diagram for a power plant coal processing system—complete with all IN/OUT interfaces. This diagram, in addition to its value in helping us to visualize the system functional structure, will identify smaller subsystem packages for use in Steps 4 to 7 if the system itself appears to be too complex to analyze in one bite. Frequently, this makes the analysis process less cumbersome, and even provides a logical basis for separating the work if more than one analyst is assigned to a given system. As a rule of thumb, we have found that systems should be represented by no more than five major functions—thus the reason for limiting the number of functional subsystems that should be used. When more than five are proposed, a close look will likely reveal that you actually have defined overlapping major functions. For example, pumping may be a part of flow regulation, not a separate functional subsystem in its own right.

The functional block diagram, in conjunction with the boundary overview (Step 2-1), provides a valuable description of the initial phase of the systems analysis process for management review. In fact, as team facilitators, we frequently suggest that this Step 3-2 be done immediately after the system boundary overview in Step 2-1 in order to reach an early consensus about whether to tackle the entire system or defer to individual subsystems to keep the analysis process manageable. For example, the system illustrated on Figure 5.9 was divided into the four subsystems shown, but only the coal feed and mill grinding subsystems received the full Classical RCM process since they were the major contributors to the system downtime problem.

5.4.3 Step 3-3—IN/OUT Interfaces

The establishment of system boundaries and the development of functional subsystems (each of which represents major functions of the system) will now permit us to complete and document the fact that a variety of elements cross the system

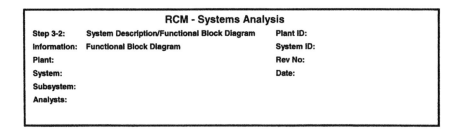

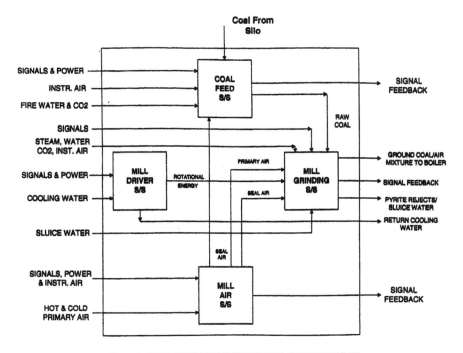

Figure 1 - COAL PROCESSING FUNCTIONAL BLOCK DIAGRAM

System:	Tuesday, April 22, 2003
Subsystem:	Page 1 of 1

Step 3-2 Functional Block Diagram
JMS Software

Figure 5.9 Typical functional block diagram on form for Step 3-2 (from Ref. 38).

boundary (or subsystem boundary if we have elected to analyze at that level). Some of the elements (electrical power, signals, heat, fluids, gases, etc.) come IN across the boundary, and some move OUT to support other systems in your plant. These are called the IN and OUT interfaces respectively, and provide additional information for us to complete the functional block diagram in Figure 5.9. This information is documented on the form shown in Figure 5.10 which, you will notice, has column headings that are identical to those of Figure 5.6. Some, but far from all, IN/OUT interfaces did occur at boundary points. But many more occur between boundary points, and these must now be identified and documented on Figure 5.10. For example, returning to our hypothetical system represented by a conference room, the boundary locations are situated in each of the eight corners of the room, but it is unlikely that these locations also have an IN or OUT interface. Rather, all IN/OUT interfaces occur across the walls, floor or ceiling defined by the eight corners. Elements such as power or phone lines, air conditioning inlets and returns, water lines, drain lines, etc. would all be listed on Figure 5.10, but none of them would have been listed on Figure 5.6. Real systems in your plant have similar situations. In fact, it is common to find that the list of IN/OUT interfaces greatly outnumbers the list that is needed to define the boundary details of Figure 5.6.

As we develop the list in Figure 5.10, and complete the lines and arrows in Figure 5.9, we can begin to clearly see that all of the OUT interfaces represent what the system produces, and these will become the focus of the principle to "preserve system function" which becomes Step 4 of our analysis process. Please note that in the systems analysis process we assume that all IN interfaces are always present and available when required. True, these IN interfaces are needed to make our system work, but the real product of our system is embodied in the OUT interfaces. Remember that IN interfaces here are OUT interfaces in some other system, so we are not really neglecting them, and, if need be, they can be analyzed as part of another system at a later date.

5.4.4 Step 3-4—System Work Breakdown Structure (SWBS)

SWBS is a carryover from terminology that was used in the Department of Defense applications of RCM, and is used to describe the compilation of the equipment (component) lists for each of the functional subsystems shown on the functional block diagram. Notice that this equipment list is defined at the component level of assembly (per the component definition described previously in Sec. 5.2) and is documented on the form shown in Figure 5.11. It is essential that all components within the system boundary be included on these equipment lists; failure to do so would automatically eliminate those "forgotten" components from any further PM consideration in Steps 4 to 7. A correct P&ID can be used as an excellent source of information to develop an equipment list. In older plants or facilities, however, it is recommended that a system walkdown also be performed to assure the accuracy of the list.

RCM - Systems Analysis

Step 3-3:	**System Description/Functional Block Diagram**	**Plant ID:**
Information:	**IN/OUT Interfaces**	**System ID** 00651-020304
Plant:	VKF HPA Auxiliary Plant	**Rev No:** 0
System:	JM3 Pumping System	**Date:** 3/10/98
Subsystem:	C92 Compressor System	
Analysts:	Ed Ivey, Brian Shields, Brown Limbaugh, Ronnie Skipworth, Glenn Hinchcliffe (facilitator)	

Type	Bounding System	Interface Location	Reference Drawing
OUT (Air)	93 A/B	Down stream side of V925	20-00054.18
IN (Air)	Atmosphere	Up stream side of inlet filter	20-00054.18
OUT (Air)	Atmosphere	Outlet of ducting	20-00054.18
IN (6.9kv)	Electrical Distribution	Cable connection at motor	
IN (480v)	Electrical Distribution	Cable connection at motor heaters	
IN (480v)	Electrical Distribution	Cable connection at oil reservoir heaters.	
IN (480v)	Electrical Distribution	Cable connection at pre-lube oil pump	
OUT (Dymac Signals)	Control Signals	Terminal strip to 501 system and PPC (C92 vibration signals and bearing temps)	
OUT (Signals)	Annunicator Panel	Terminal strip to compressor	
OUT (Signals)	Electrical Distribution	Breaker close signal for main motor and pre-lube pump	
IN (Air)	Atmosphere	Air side of motor filter	
OUT (Air)	Atmosphere	Discharge side of motor filter	
IN (Raw Water)	Raw Water System	Inlet side of valve supplying water to compressor and lube oil cooler	
OUT (Raw Water)	Raw Water System	Discharge side of raw water return valve	

System:	JM3 Pumping System	Sunday, June 08, 2003
Subsystem:	C92 Compressor System	Page 1 of 2

Step 3-3 IN/OUT Interfaces
JMS Software

Figure 5.10 Typical IN/OUT interfaces on form for Step 3-3 (courtesy of USAF/AEDC).

RCM - Systems Analysis

Step 3-4:	System Description/Functional Block Diagram	Plant ID:	
Information:	System Work Breakdown Structure	System ID	00651-020304
Plant:	VKF HPA Auxiliary Plant	Rev No:	0
System:	JM3 Pumping System	Date:	3/10/98
Subsystem:	C92 Compressor System		
Analysts:	Ed Ivey, Brian Shields, Brown Limbaugh, Ronnie Skipworth, Glenn Hinchcliffe (facilitator)		

Comp	Component ID	Component Description	Type	Qty	Dwg or Ref
01		6.9 kV Motor	Non-Instr		
02		Compressor Coupling	Non-Instr		
03		Compressor	Non-Instr		
04		Wiring and Connections	Non-Instr		
47		Gauges (press,temp) Cool Wtr: Return Press & Return, Main wtr cooling supp press/ temp, LO cooler in, Intake Vac	S - Instr		
48		Pressure/Temperature readouts (Chessel Recorder) Process Air temp by stage, Discharge air temp by stage	S - Instr		
49		Lube Oil Cooling Temperature Out	C - Instr		
50		Amp Meter for Motor Current	P - Instr		
51		Oil Reservoir Level Switch (Alarm and Trip)	P - Instr		
52		I/P Converters and Sensors (Air Operated Valves)	C - Instr		
53		Positioners and Sensors (Air Operated Valves)	C - Instr		
54		AMETEK controller for V921 and V928	C - Instr		
55		Electronic controllers for water cooling valves and V925	C - Instr		
56		Main Motor Bearing Temperature Monitors (Separate alarms and trips)	P - Instr		
57		C92 Main water cooling supply valve	Non-Instr		
58		C92 Main water cooling supply Y-strainer	Non-Instr		
59		Motor bearing lube oil valves (East and West)	Non-Instr		
60		Condensate blowdown valves (1st, 2nd, 3rd stage)	Non-Instr		

System:	JM3 Pumping System		Saturday, May 24, 2003
Subsystem:	C92 Compressor System		Page 1 of 1
		Step 3-4 SWBS	C=Control S=Status P=Protection
		JMS Software	

Figure 5.11 Typical SWBS on form for Step 3-4 (courtesy of USAF/AEDC).

Since many systems contain a sizeable complement of instrumentation and control (I&C) devices, it may be convenient to group the I&C and non-I&C components separately on Figure 5.11. We have also tried to simplify the handling of I&C components by categorizing each device as providing (1) control, (2) protection, or (3) status information only, and so indicating this distinction with a C, P, or S on Figure 5.11. We further recommend that those devices categorized as "status information only" be dropped from any further consideration in the system analysis process, and be put on the run-to-failure list. Quite simply, we believe that such devices do not warrant the expenditure of PM resources.

5.4.5 Step 3-5—Equipment History

With the possible exception of new, state-of-the-art equipments, virtually all of the components on the SWBS have some history of prior usage and operational experience. For RCM purposes, the history of most direct interest is that associated with failures that have been experienced over the past 2 or 3 years. This failure history is usually derived from work orders that were written to perform corrective maintenance tasks. The equipment history information is recorded on the form shown in Figure 5.12. Note that the primary information that we wish to capture is the failure mode and failure cause associated with the corrective maintenance action(s), since this information will be of direct value in completing Step 5, the failure mode and effects analysis (FMEA).

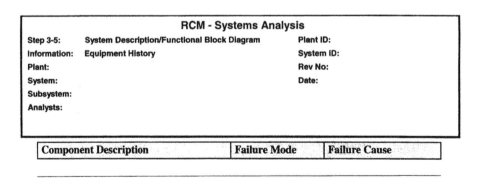

Figure 5.12 Form for equipment history, Step 3-5.

Where do we find this equipment failure history? First and foremost, if the system has already been in operation at the plant or facility in question, we should draw upon the plant-specific data that are available from the work order records or, if automated, the computerized maintenance management system (CMMS) files. In some instances, there may be sister plants or component usage in other similar facilities that are accessible from the same work order and CMMS files. Clearly, the in-house or plant-specific data are the most valuable since the records are reflective of operating and maintenance procedures that describe most accurately the actual components under investigation in the RCM analysis. In addition, there may be generic failure files that have been compiled on an industrywide basis and that contain data of considerable value on the components in question. Frequently, these generic files may not contain the identical model or drawing number of interest to you, but, with some care, components of a similar design may be applicable and useful to your analysis.

One cautionary note. The analyst should not be surprised to find that the equipment history files frequently contain very sketchy data on the failure event. Comments like "we found it broke—and fixed it" are not all that rare. This is unfortunate, not only because of its lack of useful data for your analysis, but also because it makes one wonder if the failure was ever really understood and thus fixed! Also, the data placed in the CMMS files may not be sufficient to meet the RCM analysis needs, and research back to the original work order may be necessary (if even possible at all). Even then, it is not uncommon to find a paucity of failure cause data, even when good failure mode data is present. In spite of this cautionary note, it may be appropriate to search for component failure history in support of the FMEA in Step 5. The authors' experience, however, is that this search rarely leads to useful data, so you may not wish to expend too much effort in "chasing windmills."

5.5 STEP 4—SYSTEM FUNCTIONS AND FUNCTIONAL FAILURES

The previous steps have all been directed toward developing an orderly set of information that will provide the basis for effectively pursuing the four defining features of RCM. This process begins by now defining system functions. This, of course, is done to satisfy the first RCM principle "to preserve system functions." It is therefore incumbent upon the analyst to define a complete list of system functions since subsequent steps will deal with this list in ultimately defining PM tasks that will "preserve" them. If a function is inadvertently missed, it is not likely that PM tasks directed at its preservation will be consciously considered!

It has already been noted in Step 3-3 that the development of OUT interfaces constitutes the primary source of information for identifying system functions. That is, the OUT interfaces define the system's products (a.k.a. functions). Remember, if the decision was made to perform the analysis at the subsystem level, then Steps 3 through 7 are being done at the subsystem level.

In essence, every OUT interface should be captured into a function statement. Certain OUT interfaces, however, are multiple in nature and a single function statement will suffice to cover all of them. Signals going to other plant systems or a central control room are such an example. Also, OUT interfaces represent active—and therefore readily visible—functions when Steps 2 and 3-3 have been properly done. But some functions are passive, and their subtle nature must be recognized and included by the analyst. The most obvious passive functions are structural considerations and include items such as preserving fluid boundary integrity (e.g., pipes) and structural support integrity (e.g., pipe supports).

In formulating the function statements, the analyst must keep in mind that these are not statements about what equipment is in the system. That is, avoid the use of equipment names to describe system functions. In some instances, however, reference to equipment or systems that are outside of the boundary are necessary to construct sensible function statements. Examples to illustrate correct and incorrect function statements are as follows:

Incorrect	Correct
Provide 1500 psi safety relief valves.	Provide for pressure relief above 1500 psi.
Provide a 1500 GPM centrifugal pump on the discharge side of header 26.	Maintain a flow of 1500 GPM at the outlet of header 26.
Provide alarm to control room if block valves are < 90 percent open.	Provide alarm to control room if flow rate is < 90 percent of rated value.
Provide water-cooled heat exchanger for pump lube oil.	Maintain lube oil $\leq 130°F$.

When the system functions have been defined, the analyst is ready to define the functional failures because function preservation means avoidance of functional failures. We are now embarking upon the first step in the process of determining how functions might be defeated so that we can eventually ascertain the actions to prevent, mitigate, or detect onset of function loss. We need to keep two things in mind:

1. At this stage of the analysis process, the focus is on loss of function, not loss of equipment. Thus, as with the function statements, the functional failure statements are not talking about equipment failures (this will come in Step 5).
2. Functional failures are usually more than just a single, simple statement of function loss. Most functions will have two or more loss conditions. For example, one loss condition may shut down an entire plant (a full, forced outage) while a less severe loss condition may result in only a partial forced outage or perhaps only some minor plant degradation. These distinctions are very essential so that ultimately the proper importance ranking can be determined in later portions of the analysis process (not all functional failures are equally important). In addition, these distinctions often lead to different modes of failure in the equipment that supports them, and this needs to be identified later in Step 5.

Let's illustrate this discussion with a couple of the preceding function statements.

Function	Functional failure
1. Provide for pressure relief above 1500 psi.	a. Pressure relief occurs above 1650 psi.
	b. Pressure relief occurs prematurely (below 1500 psi).
2. Maintain a flow of 1500 GPM at the outlet of header 26.	a. Flow exceeds 1500 GPM.
	b. Flow is less than 1500 but greater than 1000 GPM.
	c. Flow is less than 1000 GPM.

In Function 1, a functional failure occurs if the pressure relief is greater than the 10% margin in the design (not precisely at 1500 psi) since the pipe would rupture, but anything below 1500 psi essentially drains the system of its fluid. In both cases, the system is totally lost. In Function 2, excessive flow might violate a system design consideration and destroy some chemical process, whereas flow as low as 1000 GPM can be tolerated with some output penalty. But if flow drops below 1000 GPM, the system must be shut down.

You have probably noticed that an accurate portrayal of functional failures relies heavily on the design parameters of the system. For example, in Function 2, it was necessary to understand that there was some allowable tolerance in the flow rate before a total shutdown and complete loss of process control occurred. This range-of-conditions situation is actually a fairly common occurrence, so the analyst must be careful to ensure that the functional failures completely describe the intended design conditions for each system function. It is rare that a function is either a go or no-go. Recall that this point was emphasized earlier (in Sec. 5.4) when dealing with the development of information for Figure 5.8, the system description form. We record the function and functional failure information on the form shown in Figure 5.13.

5.6 STEP 5—FAILURE MODE AND EFFECTS ANALYSIS

5.6.1 Functional Failure–Equipment Matrix

Step 5 now brings us to the question of which component failures have the potential to defeat our principal objective to "preserve function." This will be the first time in the systems analysis process that we directly connect the system functions and the system components by identifying specific hardware failure modes that could potentially produce unwanted functional failures. In so doing, we will satisfy Feature 2 of the RCM process.

One of the major difficulties that the authors encountered in the early RCM studies was in establishing an orderly way to link and track all of the various functional failure–component combinations that required evaluation in Step 5. After several trial and error attempts, we developed the use of the functional failure–equipment

---

RCM - Systems Analysis

Step 4:	Functions/Functional Failures	**Plant ID:**	
Information:	Functional Failure Description	**System ID**	00651-020304
Plant:	VKF HPA Auxiliary Plant	**Rev No:**	0
System:	JM3 Pumping System	**Date:**	3/10/98
Subsystem:	C92 Compressor System		
Analysts:	Ed Ivey, Brian Shields, Brown Limbaugh, Ronnie Skipworth, Glenn Hinchcliffe (facilitator)		

Function #	FF #	Function/Functional Failure Description
1.0		*Supply compressed atmospheric air at 38 PSIG and 6050 CFM to 93 A/B compressors at normal operating conditions*
	1.1	No Air Supplied
	1.2	Incorrect Air Pressure
	1.3	Air supplied at off-normal operating conditions
2.0		*Provide filtered lubrication at required temperature and pressure*
	2.1	Loss of lubrication
	2.2	Lubrication at improper temperature, pressure, and cleanliness
3.0		*Remove heat of compression*
	3.1	Can not remove heat of compression
	3.2	Incorrect removal of heat of compression (high or low)
4.0		*Provide filtered atmospheric, instrument and seal air at required conditions*
	4.1	No filtered air
	4.2	Air at incorrect conditions (high or low pressure, dirty)
5.0		*Provide appropriate signals (controlling, alarming, staus, and protection)*
	5.1	No signals provided
	5.2	False signals
6.0		*Maintain boundary integrity*
	6.1	Loss of boundary integrity

System:	JM3 Pumping System	Sunday, July 20, 2003
Subsystem:	C92 Compressor System	Page 1 of 1

Step 4   Functional Failures

JMS Software

Figure 5.13 Typical function/functional failure description on form for Step 4 (courtesy of USAF/AEDC).

matrix shown in Figure 5.14. This matrix is one of the innovative additions that the authors contributed to the systems analysis process, and we often refer to it as the "connecting tissue" between function and hardware. The vertical and horizontal elements of the matrix are the component list or SWBS from Step 3-4 and the functional failure list from Step 4, respectively. The analyst's task at this point is to identify those components which have the potential to create one or more of the functional failures, and to so indicate this by placing an "X" in each appropriate intersection box. We have found that the easiest way to work through the matrix is to go down the component list one by one, and for each component, move across the functional failures. At each juncture, ask "Could something (anything) malfunction in the component such that this functional failure might occur?". A "yes" results in an "X" which tells you that this intersection needs to be evaluated in more detail using the FMEA described below. Our experience is that this matrix process proceeds quickly and accurately if the analysts have a reasonable knowledge of the system design and/or operation characteristics. However, the analysts should not be bashful in seeking assistance from engineering and operations specialists in completing the matrix. When the matrix is completed, we will have developed a specific road map to guide us through the remainder of the systems analysis process.

An example of an actual matrix that was used for the mill grinding subsystem in Figure 5.9 is shown on Figure 5.15. Initially, this matrix had Xs in all of the blocks where numbers appear. The reason for replacing many of the Xs with these numbers is discussed in the FMEA section below.

It may surprise you to find that on rare occasions we will have an empty set in Figure 5.14. That is, an item of equipment will have no X marks at any functional failure intersection. When, and if, this occurs, we have either made a mistake in our analysis or we have discovered a component that plays absolutely no useful role in the system. (The authors have seen the latter happen twice, and action was taken to remove or block out the component.) Otherwise, all components are initially viewed as "critical" in that they can play a role in creating one or more functional failures. The only question that remains is how one should prioritize that role in a world where components will ultimately compete for PM resources. This will be answered in Step 6. But to prematurely discard any component as "noncritical" without understanding its relationship to the functions and functional failures is a very dangerous course to pursue.

5.6.2 The FMEA

Upon completion of the matrix, the analysts' job now is to perform the FMEA at every intersection with an X. Our experience has shown that the best way to approach that task is to select the one or two functional failures with the most Xs in their column, and initially complete the FMEA form in Figure 5.16 at each X. As each component in the column is completed, its failure modes should be reviewed against the other functional failures with an X to see if they may equally

RCM—Systems Analysis Process		
Step 4: Functions/functional failures		
Information: Equipment–functional failure matrix	Rev no.:	Date:
Plant:	Plant ID:	
System name:	System ID:	
Analysts:		

Functional failures

No.	Equipment (or component) name											

Figure 5.14 Typical form for equipment–functional failure matrix, Step 5-1.

EQUIPMENT (OR COMPONENT) NAME	4.1.1 GRIND COAL CORRECTLY	4.1.2 CLASSIFY COAL CORRECTLY	4.2.1 SEPARATE & REMOVE PYRITES & FOREIGN OBJECTS	4.2.2 TRANSPORT REJECTS TO SLUICE SECTION	4.3.1 NO SIGNAL OUTPUT	4.3.2 ERRONEOUS SIGNALS	4.4.1 MAJOR STRUCTURAL LEAK	4.4.2 IMPROPERLY SEALED PENETRATIONS	4.5.1 VALVES TO ISOLATE ON DEMAND	4.6.1 GET CO2, STEAM, OR WATER TO CORRECT LOCATION	4.6.2 INADEQUATE FLOW OF CO2, STEAM, OR WATER	4.6.3 ISOLATE MILL GRINDING SYSTEMS	4.6.4 INERTING/FOGGING/SWIRL SYSTEMS TO OPERATE CORRECTLY	4.6.5 FAILURE TO MAINTAIN PURGE AIR
1 TABLE, YOKE AND PYRITE PLOWS	X	4.1.1	4.1.1				4.1.1	X						4.4.2
2 ROLL WHEEL ASSEMBLY(INCLUDING SEAL AIR MANIFOLD)	X						4.1.1	4.1.1						4.1.1
3 FOG AND SWIRL, PA ELBOW, AND FIRE PROTECTION			4.6.1 4.6.2		4.6.2	4.6.2	4.6.2	X		X	X		4.6.1 4.6.2	4.6.2
4 SPRING FRAME ASSEMBLY	X							4.1.1						
5 CLASSIFIER		X												
6 BSOD (BURNER SHUT-OFF DAMPER)	X	4.1.1			4.1.1 4.5.1	4.1.1 4.5.1	X		X			4.5.1		
7 STRUCTURAL HOUSING (INCLUDING PIPING)							X	4.4.1						
8 PYRITE HOPPER ASSEMBLY			X	4.2.1			X	4.4.1				4.2.1		
9 TEMP SWITCH, TS 27F, COAL/AIR EXIT ALM@180 DEG F					4.3.2	X	4.3.2	4.3.2						
10 THERMOCOUPLE, TE 727.1F COAL/AIR EXIT DCS ALARM					4.3.2	X		4.3.2						
11 DIFFERENTIAL PRESSURE TRANS, DX 27F					4.3.2	X	4.3.2	4.3.2						
12 DIFFERENTIAL PRESSURE SWITCH DS 27F * TO BE RETIRED	X				4.1.1	4.1.1	4.1.1							
13 THERMOCOUPLE, TE 727F COAL/AIR EXIT CONTROL					4.3.2	X	4.3.2	4.3.2						

Figure 5.15 Mill grinding subsystem equipment—functional failure matrix (from Ref. 38).

RCM - Systems Analysis

Step 5-2   Failure Mode and Effects Analysis

Information:	Functional Failure #			Plant ID:
Plant:				System ID:
System:				Rev No:
Subsystem:				Date:
Analysts:				

FF#	Comp	Component Description	FM #	Failure Mode	FC #	Failure Cause	Failure Effect			
							Local	System	Plant	LTA

System:
Subsystem:

Step 5-2   Failure Mode and Effects Analysis
JMS Software

Tuesday, April 22, 2003
Page 1 of 1

Figure 5-16 FMEA form for Step 5-2.

apply, in whole or in part. The chances are very good that these failure modes will satisfy at least some of the other functional failures, thus eliminating a need to repeat the FMEA exercise at several locations in the matrix. You see how this was done on Figure 5.15 where, for example, on component #1, the FMEA performed at F.F. #4.1.1 also applied at F.F. #4.1.2, 4.2.1, and 4.4.1. However, whenever an X remains, as at F.F. #4.4.2, a separate FMEA must be performed to acquire the data that is needed for that functional failure, and it is also applied at F.F. #4.6.5.

Let's discuss the guts of the FMEA process by following the columns on Figure 5.16. First, note that the header and first column clearly identifies that the information on that page is specific to one and only one functional failure. Notice also that when we identify a specific component in the second column, we have now uniquely identified one of the "X" intersections on Figure 5.14 or 5.15. Each functional failure usually has several components with an X, so there are usually multiple sheets required to complete the FMEA analysis for each functional failure. Next, we proceed to the Failure Mode column where the analysts must establish what is wrong with the component that could produce the functional failure. Usually, we may hypothesize several failure modes, but we limit our analysis to include only dominant failure modes. Dominant failure modes impose two practical restrictions on our creative ability to dissect just what might go wrong with a component:

1. In the preventive maintenance context, the failure mode must depict a problem that can be realistically addressed with a PM task. For example, we would never do PM, per se, on a microchip.
2. The failure mode must not depict an implausible situation. Examples might be:
   • inadvertent mechanical closure of a manual gate valve that is chained in the open position,
   • structural collapse of ductwork in a benign environment.

We also concern ourselves with the probability that a dominant failure mode might occur. In the early days of RCM, we saw several people try to put a threshold numerical value on this parameter, but the absence of credible failure rate data at the failure mode level made this an impossible chore. So we have successfully used the following rule: if the analysts feel that a hypothesized failure mode is likely to be seen at least once in the lifetime of the plant, then it is included. But if not, we call it a "rare event," record it for future reference, but drop it from any further use in the analysis. Our experience with this approach has been good—but not perfect. That is, we have seen "rare events" actually occur a few times after we dropped them from our analysis. When this occurs, the Living RCM Program (see Chapter 10) will catch it and correct the possible need for a PM task.

Should the sole cause of a hypothesized failure mode be human error on the part of production, operations or maintenance, we drop this from the analysis. The reason being that we cannot create a PM task to address such situations (although an IOI may be in order – see Sec. 5.10). For example, a fork lift truck drops a large pump (truck operator error) and cracks the casting into two sections (failure mode).

Please note that most components will likely have more than one failure mode associated with any given functional failure, and we should strive to identify all such dominant failure modes—both those that have already occurred and those which we believe might yet occur. We cannot emphasize too strongly the necessity to make the failure mode list as complete as possible. When we finally turn to the question of defining the PM tasks that are needed, these decisions will be irrevocably linked to these failure modes. So, no failure mode, no PM task. Failure modes are generally described in four words or less, and Figure 5.17 presents a partial listing of commonly used words to describe failure modes.

In the next column on Figure 5.16, we attempt to identify the <u>root cause</u> of each failure mode. The root cause refers to the basic reason for the failure mode—that is, why the failure mode occurred. The root cause can always be *directly* identified with the failure mode and component in question—as opposed to the consequential cause, which refers to some component failure elsewhere in the system which indirectly caused the failure mode. Consequential cause is of no interest in a maintenance analysis because no amount of maintenance on component B will ever help to avoid a failure in component A. For example, a pump motor may have seized bearings as a failure mode due to lack of proper lubrication, but this is a consequential cause resulting from failure in a separate lube oil system that feeds several items of equipment. No amount of PM on the motor bearings will prevent a basic

abrasion	damaged	lack of —	ruptured
arcing	defective	leak	scored
backward	delaminated	loose	scratched
out of balance	deteriorated	lost	separated
bent	disconnected	melted	shattered
binding	dirty	missing	sheared
blown	disintegrated	nicked	shorted
broken	ductile	notched	split
buckled	embrittlement	open	sticking
burned	eroded	overheat	torn
chafed	exploded	overtemp	twisted
chipped	false indication	overload	unbonded
clogged	fatigue	overstress	unstable
collapsed	fluctuates	overpressure	warped
cut	frayed	overspeed	worn
contaminated	intermittent	pitted	
corroded	incorrect	plugged	
cracked	jammed	punctured	

Figure 5.17 Typical descriptors for failure modes.

failure mode of, say, a clogged filter in the separate lube oil system. However, bearing seizure due to contamination buildup in a self-contained oil reserve is a root cause that can only be addressed at the motor itself. The analyst will undoubtedly find that equipment history files are quite sparse when it comes to root cause information. This again points to a glaring deficiency in data systems. (If you don't know why a failure occurred, how can you be sure it's fixed?) But this is a fact of life, and the author's advice here is to do your best to intelligently select one or two likely root causes for entry onto the form. The reason for our emphasis on attempting to establish a root cause, even if only "guesstimated," is that this piece of input may eventually prove crucial in selecting a candidate PM task. In addition, some (less than 5%) failure modes have two credible root causes for a single failure mode, and could conceivably require two different PM tasks for the same failure mode!

The final step in the FMEA process is the effects analysis portion of the form. Here, the analyst will determine the consequence of the failure mode, and this will be done at three levels of consideration—locally, at the level of the component in question; at the system level; and finally, at the plant level. There are two primary reasons for conducting the effects analysis at this point: (1) we want to assure ourselves that the failure mode in question does, in fact, have a potential relationship to the functional failure being studied; and (2) we want to introduce an initial screening of failure modes that, by themselves, cannot lead to a detrimental system or plant consequence. In order to fully understand the significance of these two statements, we need to introduce and discuss the single failure rule and how we treat redundancies that have been designed into the system.

5.6.3 Redundancy—General Rule

Our objective in RCM is to preserve function. Thus, in the maintenance strategy of how we view the commitment of resources, it becomes important to first commit those resources to single failure occurrences that detrimentally impact function (i.e., we do not consider multiple failure scenarios as is frequently done in safety analyses). If redundancy prevents loss of function, then a failure mode thus shielded by redundancy should not be given the same priority or stature of a failure mode that can singly defeat a necessary function. Note that if one is truly concerned that there is a high probability of multiple independent failures in a redundant configuration, then what you have identified is a more fundamental design issue, and not one that should be addressed or solved by the maintenance program.

So, how do we invoke this redundancy rule? Quite simply, when listing the failure modes, we do not introduce the redundancy rule since our objective is to assure that we initially capture each failure mode (protected or not) that can lead to the functional failure. But then, in the effects analysis, we apply the redundancy rule. If available redundancy essentially eliminates any effect at the system level (and, it will follow, at the plant level also), we drop the failure mode from further consideration, and place it on the run-to-failure (RTF) list that will receive a further review and sanity check in Step 7. Since complex plants and facilities are often designed with a host of redundancy features in order to achieve high levels of safety and productivity, it is not uncommon to find that this initial screening with the redundancy rule could relegate 50% or more of the failure modes to the RTF status. Should you encounter this situation, your maintenance program will likely realize significant cost reductions from the foresight that occurred during the design phase (even though that foresight was not, in all likelihood, maintenance driven).

5.6.4 Redundancy—Alarm and Protection Logic

There is one important exception to the preceding rule, which involves alarms, inhibits or permissives, isolation and protection logic devices involving some voting scheme. Here, the rule requires an assumption of multiple failures in order to properly assess the effects or consequences of redundancy loss. In the case of alarms (as well as isolation, inhibit, and permissive devices), a "failure to operate" is, by itself, not significant. It can become significant, however, if the alarmed or protected component has also failed. So, we assume that the alarmed component has failed in order to place the proper perspective in the effects analysis on the consequence of not knowing that such has occurred. The same principle holds with protection logic, where redundant channels are assumed failed to the extent that the next single failure will wipe out the protection logic. We tend to find protection logic systems when dealing with safety and environmental issues or areas where "trips" must occur automatically to preclude widespread damage to a plant.

Again, if the system effect (and thus the plant effect) is "none" as a result of applying the redundancy rule, we drop the failure mode from further consideration in the analysis until the final sanity check (this occurs in Step 7-2). Conversely, when there is some form of system and/or plant effect, we retain the failure mode for further consideration. When there is a choice to be made from several possible failure effects, we always choose the worst-case scenario in order to reflect the most severe consequence that could result from the failure mode. Such choices could occur, for example, as a function of time or occurrence (start-up, steady state, etc.) or plant operating parameters (flow rate, pressure, temperature, etc.). The last column on the FMEA form, labeled LTA or logic tree analysis, is where we signify a yes or no to indicate whether or not the failure mode will be carried forth to Step 6—the logic (decision) tree analysis (LTA). The Step 5 process is continued until each functional failure and its related components have been through the FMEA. A typical completed FMEA sheet is shown on Figure 5.18.

5.7 STEP 6—LOGIC (DECISION) TREE ANALYSIS (LTA)

The failure modes that survive the initial screening test in the effects analysis in Step 5 will now be further classified in a qualitative process known as logic

FF#	Comp	Component Description	FM #	Failure Mode	FC #	Failure Cause	Local	Failure Effect		
								System	Plant	LTA
1.1	03	Compressor	3.26	Air cooler fins debond from tube	3.26.1	Normal use and wear	Loss of air cooling efficiency resulting in increase inlet air temperature to next stage which can cause damage.	Worst case- Can't supply air to C93A/B	Loss of high pressure air (HPA)	YES
1.1	03	Compressor	3.27	Air coolers buildup of dirt on water side	3.27.1	Dirty water	Loss of air cooling efficiency	Worst case- Can't supply air to C93A/B	Loss of high pressure air (HPA	YES
1.1	03	Compressor	3.28	Moisture separator gets dirty	3.28.1	Normal use and wear	Restricts air flow resulting in possible surge	Worst case- Can't supply air to C93A/B	Loss of high pressure air (HPA	YES
1.1	03	Compressor	3.29	Inlet/outlet piping gasket	3.29.1	Aging	Air leak to atmosphere causing loss of compressor efficiency	Reduced air supply to C93A/B	Gradual loss of ability to produce HPA	YES
1.1	03	Compressor	3.30	Cooler waffle gasket deterioration or debonds	3.30.1	Age	Improper water flow resulting in lack of air cooling	Worst case- Can't supply air to C93A/B	Loss of high pressure air (HPA)	YES
1.1	04	Wiring and Connections	4.01	Insulation failure leading to a short	4.1.1	Aging and heat	Locally will see cracked or bare wire	Worst case- Can't supply air to C93A/B	Loss of high pressure air (HPA)	YES
1.1	04	Wiring and Connections	4.02	Connections become loose, broken, or corroded	4.2.1	Vibration and moisture	I/C- false readings Power- localized heating	Worst case- Can't supply air to C93A/B	Loss of high pressure air (HPA)	YES
1.1	05	Process Air Relief Valve LPA-RV-921-U	5.01	Loss of spring tension	5.1.1	Corrosion and use	Inadvertent actuation	Can't supply air to C93A/B	Loss of high pressure air (HPA)	YES
1.1	05	Process Air Relief Valve LPA-RV-921-U	5.02	Valve sticks closed	5.2.1	Corrosion and insufficient use	Loss of protection	Compressor will go into surge and can't supply air to C93.	Loss of high pressure air (HPA)	YES
1.1	05	Process Air Relief Valve LPA-RV-921-U	5.03	Fails to reseat - RARE EVENT	5.3.1	Considered to be a rare event			Loss of high pressure air (HPA)	YES
1.1	06	V925A Check Valve	6.01	Fails to reseat	6.1.1	1) Stuck disk, broken hinge pin, and bent disk arm 2) Dirt and debris	Allows air to flow in wrong direction	Compressor damage	Loss of high pressure air (HPA)	YES

System: JM3 Pumping System
Subsystem: C92 Compressor System

Step 5-2 Failure Mode and Effects Analysis
JMS Software

Tuesday, April 22, 2003
Page 5 of 23

Figure 5.18 Typical FMEA on form for Step 5-2 (courtesy of USAF/AEDC).

tree or decision tree analysis (LTA). The purpose of this step is to further prioritize the emphasis and resources that should be devoted to each failure mode, recognizing as we have earlier that all functions, functional failures and, hence, failure modes are not created equal. Thus Step 6 satisfies Feature 3 of the RCM process.

Several ranking schemes could conceivably be used to achieve a priority listing of the failure modes, but the RCM process uses a simple three-question logic or decision structure that permits the analyst to quickly and accurately place each failure mode into one of four categories (or bins as we often call them). Each question is answered either yes or no. As we shall see momentarily, the bins form a natural importance ordering to the failure modes.

The basic LTA uses the decision tree structure shown in Figure 5.19. The information that is gathered from this tree is recorded on the form shown in Figure 5.20. You will notice that this decision process will identify each failure mode in one of three distinct bins: (1) safety-related, (2) outage-related, or (3) economics-related. It also distinguishes between evident (to the operator) or hidden. Let's examine the details of how the LTA is used.

Each failure mode is entered into the top box of the tree on Figure 5.19, where the first question is posed: Does the operator, in the normal course of his or her duties, know that something of an abnormal or detrimental nature has occurred in the plant? It is not necessary that the operator should know exactly what is awry for the answer to be yes. The reason for this question is to establish initially those failure modes that may be hidden from the operator. Failures in standby systems or components are typical of hidden failures; unless some deliberate action is taken to find them, they will not be discovered until a demand is made, and then it may be too late. Thus, hidden failures could later give rise to failure-finding PM tasks. Evident failures, however, alert the operators to act, including taking the necessary steps to detect and isolate the failure mode if such is not immediately visible. So a yes to the first question leads us to the next question in the tree while a no leads us directly to bin D—or the hidden function bin.

All failure modes, both evident and hidden, now pass to the second question, which asks if they can lead to a safety problem. Safety, in the context used here, refers to personnel death or injury, either on-site or off-site. However, you can define safety in whatever fashion your particular needs may dictate. For example, safety may be limited to include only off-site injuries or deaths; or safety may be defined to include violation of EPA standards or even equipment damage. The authors' preference is to limit safety to personnel injury or death, but this is strictly a personal choice. When broadened beyond this, there may be different levels of safety that must be classified. In any event, if the second question yields a yes, the failure mode is placed in bin A—or the safety bin. A "no" takes us to the third and final question.

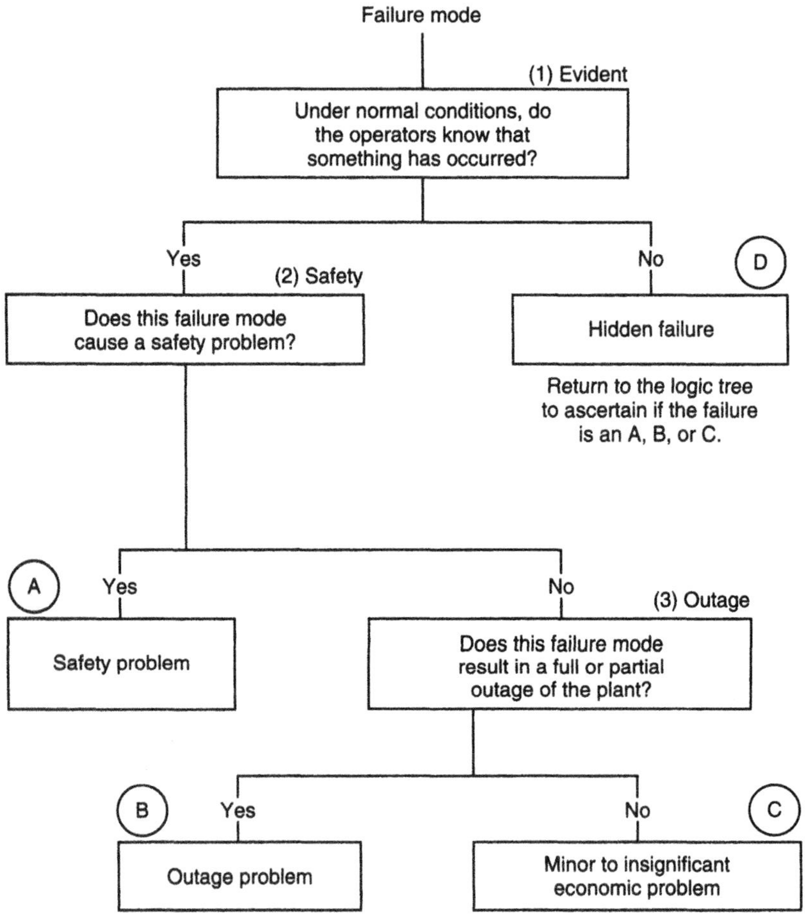

Figure 5.19 Logic tree analysis structure.

If there is no safety issue involved, the remaining consequence of interest deals solely with plant or facility economics. Thus, the third question is formulated to make a simple split between a large (and usually intolerable) economic penalty, and a lesser (and usually tolerable for at least some finite time period) economic penalty. This is done by focusing on plant outage or loss of productivity. The question becomes: Does the failure mode result in a loss of output >5%? This can also be stated as: Does the failure mode result in a full or partial plant outage (where partial can be defined as > 5%)? The selection of the 5% threshold value depends upon several variables, so the analyst should adjust this value to suit the situation at hand. A "yes" answer puts us in bin B, which is the outage bin, and

RCM - Systems Analysis

Step 6	**Logic Tree Analysis**	**Plant ID:**
Information:	**Failure Mode Critcality**	**System ID** 00651-020304
Plant:	VKF HPA Auxiliary Plant	**Rev No:**
System:	JM3 Pumping System	**Date:**
Subsystem:	C92 Compressor System	
Analysts:	Ed Ivey, Brian Shields, Brown Limbaugh, Ronnie Skipworth, Glenn Hinchcliffe (facilitator)	

FF #	Comp #	Component Description	FM #	Failure Mode	Evident?	Safety	Outage	Cat	Comments
1.1	03	Compressor	3.29	Inlet/outlet piping gasket	YES	NO	YES	B	
1.1	03	Compressor	3.30	Cooler waffle gasket deterioration or debonds	NO	NO	YES	D/B	
1.1	04	Wiring and Connections	4.01	Insulation failure leading to a short	NO	YES	YES	D/A	
1.1	04	Wiring and Connections	4.02	Connections become loose, broken, or corroded	NO	NO	YES	D/B	
1.1	05	Process Air Relief Valve LPA-RV-921-U	5.01	Loss of spring tension	NO	YES	YES	D/A	
1.1	05	Process Air Relief Valve LPA-RV-921-U	5.02	Valve sticks closed	NO	YES	YES	D/A	Valve has never been bench tested
1.1	06	V925A Check Valve	6.01	Fails to reseat	NO	NO	YES	D/B	
1.1	06	V925A Check Valve	6.02	Fails to open	YES	NO	YES	B	
1.1	07	V921 Air Operated Control Valve	7.01	Packing leak	NO	NO	NO	D/C	
1.1	07	V921 Air Operated Control Valve	7.02	Bound stem	YES	NO	YES	B	

System:	JM3 Pumping System	
Subsystem:	C92 Compressor System	

Step 6   Logic Tree Analysis
JMS Software

Saturday, May 24, 2003

Page 1 of 1

Figure 5.20 Typical logic tree analysis on form for Step 6 (courtesy of USAF/AEDC).

signifies a significant loss of income as the consequence. As an example, a full outage in a base-load 800 MWe electric power plant can cost upwards of $750,000 per day to purchase replacement power! A "no" answer tells us that the economic loss is small and places us in bin C. That is, the failure mode is essentially tolerable until the next target of opportunity arises to restore the equipment to full specification performance. There are many examples of bin-C-type failure modes, and these include items such as small leaks and degraded heat transfer due to tube scaling.

When the LTA process is concluded, every failure mode that was passed to the LTA will have been classified as either A, B, C, D/A, D/B, or D/C.

What do we do with this information? We use it, if you will, to separate the wheat from the chaff. In a world of finite resources, in other words, who gets the favorable nod? I am sure, by now, that failure modes, either evident or hidden, which land in bin A or bin B would have your priority over bin C. And, in general, bin A has priority over bin B. The current litigation environment, if nothing else, makes that choice an easy one. So, we usually choose to address PM priorities as:

1. A or D/A
2. B or D/B
3. C or D/C

Bin C, in particular, tends to raise a dilemma in the sense that its potential consequence is small, by definition—but we hate to just walk away from those failure modes. While each case must be viewed on its own merits, we should note that the evidence is rather strong that bin C should be relegated to the run-to-failure list without further ado. If this should be done, you will notice that the accumulation of RTFs from the "status only" instruments, the effects analysis in Step 5, and now the RTFs from bin C in Step 6 could sum to a sizeable list! In most instances, the sanity check on these failure modes will leave them in their RTF status. This list alone, at this point in the systems analysis process, could constitute a sizable reduction in O&M costs if all current PM tasks which apply to these failure modes were eliminated! It is the authors' recommendation that all bin C failure modes be designated as RTF, and changed only if they should not pass the sanity check in Step 7-2. In this case, only failure modes that have been assigned an A or B classification would be passed on to Step 7-1.

5.8 STEP 7—TASK SELECTION

5.8.1 Step 7-1—The Task Selection Process

Our systems analysis efforts to this point have been directed to delineating those failure modes where a PM task will give us the biggest return for the investment to be made. The LTA results in Step 6 have specifically identified those failure modes with the designation of A, D/A, B, or D/B. So, for each of these failure modes, our job now is to determine the list of applicable candidate tasks, and then to select the most effective task from among the competing candidates. Recall from Sec. 4.4 that PM task selection in the RCM process requires that each task meet the applicable and effective test, which is defined as follows:

- *Applicable*. The task will prevent or mitigate failure, detect onset of failure, or discover a hidden failure.
- *Effective*. The task is the most cost-effective option among the competing candidates.

If no applicable task exists, then the only option is RTF. Likewise, if the cost of an applicable PM task exceeds the cumulative costs associated with failure, then the effective task option will also be RTF. The exception to this rule would be a bin A, or safety-related, failure mode where a design modification may be mandatory.

Developing the candidate list of PM tasks is a crucial step, and frequently requires help from several sources. Again, involvement in task selection from the plant maintenance personnel is necessary to realize the benefit of their experience as well as to gain their buy-in to the RCM process. However, other sources of input—such as operations personnel, technical data searches, and vendor expert advice—are recommended to assure the inclusion of state-of-the-art technology and techniques. This latter statement is especially true regarding the introduction of performance monitoring and predictive maintenance options for CD tasks.

The road map in Figure 5.21 and the form in Figure 5.22 are used to structure and record the task selection process. The road map, in particular, is very useful in helping the analyst to logically develop the candidate PM tasks for each failure mode. Briefly, the steps in Figure 5.21 are as follows:

1. We have previously discussed the significance associated with a knowledge of an equipment's failure density function (Sec. 3.4) and the danger that befalls us should we erroneously select overhaul (or intrusive TD) tasks if we do not know the failure density function (Sec. 4.2). Thus, this first question requires that we acknowledge how much we really know about the equipment age–reliability relationship (i.e., the failure density function). In all likelihood, we rarely know it precisely but, occasionally, we may have some reasonable estimate of it and can answer yes. Or more likely, we may have some information about failure cause (from the FMEA data on Figure 5.16) that indicates aging or wearout (i.e., the back end of the bathtub curve), and can give a partial yes to question (1) even though we do not know when the failure mode

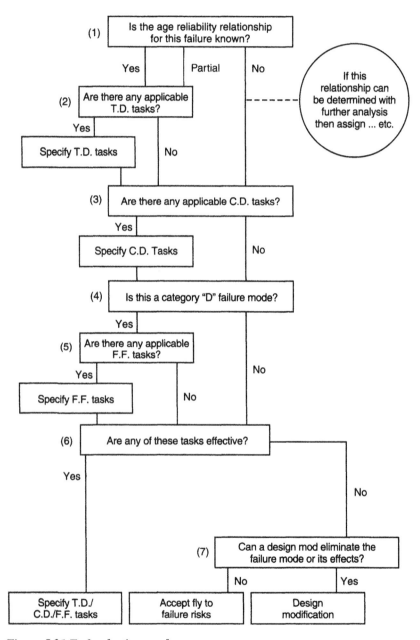

Figure 5.21 Task selection road map.

RCM - Systems Analysis

Step 7-1   Task Selection   Plant ID:

Information: Selection Process and Decision   System ID:

Plant: Rev No:

System: Date:

Subsystem:

Analysts:

FF #	Comp	Comp Desc	FM #	Failure Mode	FC #	Failure Cause	1	2	3	4	5	6	7	Candidate Task	Effective Info	Selective Dec.	Est Freq

System: Step 7-1   Task Selection   Tuesday, April 22, 2003

Subsystem: JMS Software   Page 1 of 1

Figure 5.22 Task selection form for Step 7-1.

might initiate. But in the absence of at least a reasonable estimate, we must answer no, and not fall into the trap of trying to search for an applicable TD task which could actually make things worse—not better.

2. When we do have the age–reliability information (or least some partial evidence of such), this signals that we do understand (or can conservatively estimate) the mechanisms/causes associated with the failure mode, and possibly how the failure rate deteriorates over time. In other words, we know what TD task to select to prevent or mitigate the failure mode and also approximately when it should be done to minimize the chance of its occurrence. Note that if the age–reliability information shows a constant failure rate over the total expected life of the equipment, there is no applicable TD task available to us because any failure occurrence is strictly random in nature.

3. Even if a candidate TD task has been defined, we will further explore the possibility of an applicable CD task(s). Frequently, this is a smart path to pursue if our age–reliability information (and thus TD task selection) is on somewhat shaky ground. An appropriate CD task, aimed at measuring some telltale parameter over time, may well be the best selection ultimately. It will help us to develop the age–reliability information and, until then, will give us a high confidence in taking preventive measures (usually overhaul in nature) at the correct time. Of course, if your plant mimics the data in Figure 4.1, the CD task may well tell you that you never see the onset of the failure mode during the useful life of the equipment! Hopefully, if the answer to question (1) was no, you will find at least one candidate task in question (3). Don't be surprised, however, if you do not find either a TD or CD candidate task, since some failure modes are just not amenable to a PM action, even when age–reliability information is fully known.

4. Going back to the LTA information, is this a hidden failure mode?

5. If yes to question (4), can we specify a candidate non-intrusive FF task? In all likelihood, there will be some type of failure finding that can be considered. It is rare that some form of failure finding test or inspection cannot be done. When an FF task is selected, we would define its frequency such as to eliminate or significantly minimize any system or plant downtime that might be required to correct the failure.

6. Now we are ready to examine the relative costs associated with each candidate task, and this always includes the option of RTF costs. The job here is to select the lowest cost option. You may wish to refer to Sec. 2.3 where the discussion on failure finding, using the auto spare tire example, illustrates how it is possible to have applicable TD, CD, and FF tasks with the ultimate selection being based on the effectiveness measure.

7. This question is aimed at directing the analyst to consider a design modification as a solution when no applicable and effective task has

been identified. In the case of bin A or safety-related failure modes, consideration of a design modification should be mandatory and presented to management for final decision on what should be done. Otherwise the default choice is run-to-failure (RTF). The form in Figure 5.22 is used to record all of the decisions that were made during the task selection process, including the final selection which is recorded in the "Selective Dec." column. The last column, "Est. Freq.," is where we will record the suggested frequency or interval that should be assigned to the task. Some further remarks on the subject of task periodicity may be found in Sec. 5.9. A completed sample of the task selection is shown in Figure 5.23.

5.8.2 Step 7-2—Sanity Check

At key points throughout the systems analysis process, we have been collecting components and failure modes on an RTF list:

- On the SWBS, instruments were for "status information only."
- In the FMEA, their effects were local only.
- In the LTA, they were prioritized as Bin C or D/C.

The form on which we could collect such a list is shown in Figure 5.24, which is now used to complete the sanity checklist process. The basis for such a sanity check derives from the possibility that there are valid reasons for performing a PM task even though the failure mode is not directly or solely related to a high-priority function. Referring to Figure 5.24, we see eight reasons listed on the form, and each individual situation may generate other reasons which must be considered. The eight so listed have been drawn from the authors' experience, and are as follows:

1. *Marginal effectiveness.* It is not totally clear that the RTF costs are significantly less than the PM costs.
2. *High-cost failure.* While there is no loss of a critical function, the failure mode is likely to cause such extensive and costly damage to the component that it should be avoided.
3. *Secondary damage.* Similar to item 2, except that there is a high probability that the failure mode could lead to extensive damage in neighboring components, and possibly loss of critical functions due to the domino effect.
4. *OEM conflict.* The original equipment manufacturer recommends a PM task that is not supported by the RCM process. This dichotomy is especially sensitive if warranty conditions are involved.
5. *Internal conflict.* Maintenance or operations personnel feel strongly about a PM task that is not supported by the RCM process. While these feelings can be more emotional than technical in their basis, management may decide against the RCM finding.

RCM - Systems Analysis

Step 7-1	**Task Selection**	Plant ID:
Information:	**Selection Process and Decision**	System ID 00651-020304
Plant:	VKF HPA Auxiliary Plant	Rev No:
System:	JM3 Pumping System	Date:
Subsystem:	C92 Compressor System	
Analysts:	Ed Ivey, Brian Shields, Brown Limbaugh, Ronnie Skipworth, Glenn Hinchcliffe (facilitator)	

FF #	Comp	Comp Desc	FM #	Failure Mode	FC #	Failure Cause	1	2	3	4	5	6	7	Candidate Task	Effective Info	Selective Dec.	Est Freq
1.1	03	Compressor	3.30	Cooler waffle gasket deterioration or debonds	3.30.1	Age	N	Y	N	Y	N	N	-	1. Tear down and inspect-TDI 2. Visually check for water leak-TD 3. RTF		Visually check for water leak-TD	1/ Shift
1.1	04	Wiring and Connections	4.01	Insulation failure leading to a short	4.1.1	Aging and heat	P	Y	N	Y	N	N	-	1. Inspect and Megger wiring - TDI 2. Inspect exposed wiring for damage when performing Calibration and other PM's - TD 3. RTF		Inspect exposed wiring for damage and deterioration when performing Calibration and other PM's - TD	To be done during applicable calibration tasks (6 Mo. and 1 Yr.)
1.1	05	Process Air Relief Valve LPA-RV-921-U	5.02	Valve sticks closed	5.2.1	Corrosion and insufficient use	P	Y	N	Y	N	Y	-	1. Periodically replace spring or valve - TDI 2. Remove and bench test - TDI 3. Manually lift test by unseating the valve - TDI 4. RTF	Perform initial calibration for baseline	Remove and bench test - TDI	5 Yr.

System:	JM3 Pumping System	**Step 7-1 Task Selection**	Saturday, May 24, 2003
Subsystem:	C92 Compressor System	**JMS Software**	Page 1 of 1

Figure 5.23 Typical task selection on form for Step 7-1 (courtesy of USAF/AEDC).

Figure 5.24 Sanity checklist for Step 7-2.

6. *Regulatory conflict.* Stipulations by a regulatory body (e.g., Nuclear Regulatory Commission, EPA) have established a PM task that is not supported by the RCM process. Should the RCM finding be argued with the regulators?

7. *Insurance conflict.* Similar to preceding items 4 and 6, and thus following the RCM finding would necessitate a change in the agreement with the insurance company.

8. *Hidden.* A re-evaluation of a failure mode categorized as D/C where it may not be prudent to permit a hidden failure to ever reach a full failure state.

The final selection decision is then indicated in the column so labeled. A "yes" in one of the above eight columns does not automatically denote that the RTF status is rejected, but in most cases experience says that the analysts will reject RTF in favor of some reasonable PM task. When this latter course is chosen, the task selection and its frequency are made on the spot and so recorded. If the original RTF status is upheld, then the selection decision will so denote by indicating RTF. A completed sample of the sanity check is shown in Figure 5.25.

5.8.3 Step 7-3—Task Comparison

If your RCM application is to an existing plant or facility, then there is a PM program of one sort or another already in place. One of several issues may be motivating management to upgrade this existing program. But, in so doing, it is fairly certain that management will want to know how the RCM-based PM tasks stack up against the current PM tasks. How different is the RCM program, and what is the nature of those differences? Even in a new plant or facility, it may be very important to compare the OEM recommendations with the RCM-based PM tasks. The form in Figure 5.26 is used to collect such comparison information.

The selected PM tasks from Step 7-1 (Figure 5.22) and Step 7-2 (Figure 5.24) are listed in the "RCM Selection Dec/Cat" and "Est. Freq." columns in Figure 5.26 and their components and failure modes of origin are likewise listed for traceability purposes. A completed sample of the task comparison form is shown in Figure 5.27. Since there is probably no established PM task–failure mode relationship in the existing program, the analyst can only try to match current tasks with the RCM tasks to see where they may be alike. Any task that does not match is then listed in the "Current Task Description" column with no counterpart RCM task listed at all. When the form is complete, it will contain four distinct categories of comparison:

1. RCM-based and current PM tasks are identical.
2. Current PM tasks exist, but should be modified to meet the RCM-based tasks.
3. RCM-based PM tasks are recommended where no current tasks exist.

RCM - Systems Analysis

Step 7-2:	Task Selection
Information:	Sanity Checklist
Plant:	VKF HPA Auxiliary Plant
System:	JM3 Pumping System
Subsystem:	C92 Compressor System
Analysts:	Ed Ivey, Brian Shields, Brown Limbaugh, Ronnie Skipworth, Glenn Hinchcliffe (facilitator)

Plant ID:	
System ID:	00651-020304
Rev No:	0
Date:	7/14/98

FF #	Comp #	Comp Desc	FM #	Failure Mode	Marg Eff	High Cost	Sec Dmg	OEM Conf	Int Conf	Reg Conf	Insur Conf	Hidden	RTF	Selection Dec	Est Freq	Comments
1.1	01	6.9 kV Motor	1.05	RTD fails open	NO	NO	NO	NO	NO	NO	NO	NO	YES	RTF		
1.1	01	6.9 kV Motor	1.06	Air filter clogging	NO	NO	NO	NO	NO	NO	NO	YES	NO	Periodically replace filters - TDI	1M w/ AE	See IOI #9
1.1	03	Compressor	3.04	Oil leak	NO	NO	NO	NO	NO	NO	NO	NO	YES	RTF		
1.1	07	V921 Air Operated Control Valve	7.01	Packing leak	NO	NO	NO	NO	NO	NO	NO	YES	YES	RTF		Not considered cost effective, has little effect on operations until very large, no failure history
1.1	08	V928 Air Operated Control Valve	8.01	Packing leak	NO	NO	NO	NO	NO	NO	NO	NO	YES	RTF		No cost effective task
1.1	09	V925 Air Operated Control Valve	9.01	Packing leak	NO	NO	NO	NO	NO	NO	NO	NO	YES	RTF		No cost effective task
1.1	11	Interstage Cooler Water Control Valve (1st, 2nd, 3rd Stages)	11.01	Packing leak	YES	NO	NO	NO	NO	NO	NO	NO	NO	Operator to look for valve leaks - TD	1/ Shift	

System:	JM3 Pumping System
Subsystem:	C92 Compressor System

Step 7-2   Sanity Checklist
JMS Software

Saturday, May 24, 2003
Page 1 of 5

Figure 5.25 Typical sanity checklist on form for Step 7-2 (courtesy of USAF/AEDC).

Figure 5.26 Task comparison form for Step 7-3.

RCM - Systems Analysis

Step 7-3:	Task Selection	Plant ID:	
Information:	Comparison RCM vs Current PM Task	System ID	00651-020304
Plant:	VKF HPA Auxiliary Plant	Rev No:	
System:	JM3 Pumping System	Date:	
Subsystem:	C92 Compressor System		
Analysts:	Ed Ivey, Brian Shields, Brown Limbaugh, Ronnie Skipworth, Glenn Hinchcliffe (facilitator)		

FF#	Comp #	Comp Description	FM #	Fail Mode/Where From	RCM Selection Dec/Cat	Est Freq	Current Task Description	Freq
1.1	01	6.9 kV Motor	1.06	Air filter clogging	Periodically replace filters - TDI	1M w/ AE	Inspect cleanliness of air ducts - TDI	4 Yr.
1.1	03	Compressor	3.28	Moisture separator gets dirty	Evaluate Air Cooler performance data for abnormal trends and indications-CD	1 Mo.	None	-
					D/B			
1.1	03	Compressor	3.29	Inlet/outlet piping gasket	Perform leak cooler leak check-TD	1/ Shift	None	-
					B			
1.1	04	Wiring and Connections	4.01	Insulation failure leading to a short	Inspect exposed wiring for damage and deterioration when performing Calibration and other PM's - TD	To be done during applicable calibration tasks (6 Mo. and 1 Yr.)	Verify proper operation and calibrate - TDI Ref: MP-SD-00651-041402 MP-TD-00651-041401 MP-SD-00651-041405 MP-TP-00651-041402 NOTE: There are no specific references to wiring in these procedures; one should be added.	Ref 1) 6 Mo. Ref 2) 1 Yr. Ref 3) 1 Yr. Ref 4) 1 Yr.
					D/A			

System:	JM3 Pumping System		Saturday, May 24, 2003
Subsystem:	C92 Compressor System		Page 1 of 1

Step 7-3   Comparison RCM vs Current PM Task
JMS Software

Figure 5.27 Typical task comparison on form for Step 7-3.

4. Current PM tasks exist where no RCM-based tasks are recommended, and are therefore candidates for deletion.

These comparison categories can be further refined to produce some charts with excellent visibility for management consumption, and examples of how this might be done are shown in Sec. 12.2—Selected Case History Studies. Further, the analyst can use the fourth category as a checklist to see if any obvious failure mode or PM task was inadvertently missed in the systems analysis process.

You will recall that in Sec. 5.2 it was recommended that the analyst defer the gathering of current PM task data during the information collection part of Step 1 until this point in Step 7 was reached. Hopefully, you can now appreciate the basis for this recommendation; namely, to avoid biasing any portion of the RCM process with existing practices, so that the comparison process done here could truly be viewed as two independent paths for defining a PM program for the same system.

When the task comparison has been completed, the systems analysis process is essentially complete. At this point in the process, the task comparison information represents the summary listing of the RCM findings for the system in question. Note that most components will have more than one failure mode identified throughout the course of the FMEA exercise. Thus, the analyst must be careful to recognize that all failure modes must result in the RTF decision before the component itself can be declared as RTF. This caution, of course, does not deter the analyst from deciding that some component failure modes are RTF, while others require some PM action.

Each organization has its own culture with respect to a management review and approval process. As a general rule, Figure 5.26 represents very good summary-level information for this purpose. At a minimum, the plant or facility manager and the maintenance and operations supervisor on his or her staff should be required to approve these results before implementation is initiated. A book containing the completed systems analysis information (i.e., the forms developed for Steps 1 to 7) should be available as a backup in order to show the details of the "how and why" behind each specific finding. Challenges to the findings are to be expected, and if the analyst has done the necessary homework, the conclusions and findings should be self-evident. But the process is not perfect, and adjustments during the management review are a constructive part of any RCM program.

5.9 TASK INTERVAL AND AGE EXPLORATION

Selection of the correct interval (or frequency or periodicity) at which to perform a preventive maintenance task is, by far, the most difficult job confronting the maintenance technician and analyst. We have just described a very systematic and credible method to select "what" PM tasks should be done—i.e., the RCM systems analysis process. But nowhere in that process did we directly spell out just "when" those tasks are to be performed. Determination of the task interval is a very difficult problem, mainly because it is associated with a very elusive parameter—time (or some equivalent thereof such as cycles, miles, etc.). More precisely, we need to understand how physical processes and materials change over time, and how those changes ultimately lead to what we call failure modes. So, in reality, we are dealing with failure rates and the need to know how these failure rates can vary as a function of time. Does this begin to ring a familiar bell? It should if you recall what we discussed in Secs. 3.4 and 4.2. We are actually entering into the world of statistical analysis in order to tackle a solution. But we will try to keep it as simple as possible, and in the process hopefully to offer a viable approach to this necessary piece of information.

When the task selection process has employed the road map presented in Figure 5.21, we will at least have established at the outset whether we know the age–reliability relationship for the specific failure mode in question. This relationship, ideally, is the key item of information that is needed to initially consider the practicality of seeking a TD task whose objective would be to prevent the onset of a known aging or wearout failure mechanism. Now, if we do know the age–reliability relationship, then we also have the precise information that we need to select the TD task interval. That is, we have the failure density function (fdf) for the failure mode population, and we can select the task interval from the statistical knowledge by simply deciding on the level of consumer risk that we want to accept. Suppose, for example, that the fdf looks like the bell-shaped curve shown in Figure 3.1, where the x-axis is operating time and the y-axis is probability of failure. The left-hand tail may be quite long, thus signifying an extended period of time during which the probability of failure is quite small and, for all practical purposes, the item is in a constant failure rate condition. That is, the aging/wearout failure mode has not yet really started to exhibit itself. Recall that we did indeed see such situations in the curves shown in Figure 4.1 (curves D, E, and F). However, as we proceed to the right in Figure 4.1, or as we see the probability of failure beginning to increase as additional operating time is accumulated, we can now decide just how far we want to proceed before doing the TD task. And this is where the level of consumer risk comes into play. We can pick that level of risk by selecting the percentage of area under the fdf that we can tolerate before taking action. Say we choose 15%. This means that there is a 15% chance that the failure mode could occur before we take the preventive actions. Notice that we can choose any percentage value that we want; the only question is how much risk we want to take since decreasing risk leads to more frequent PM actions and higher PM costs. Notice that if we use the mean (or MTBF) for the bell-shaped fdf, there is a 50% chance of failure before we take preventive actions. For other fdfs, the chance of failure can be as large as 67% when the mean is used. This is clearly not an acceptable level of risk in most circumstances—hence, using an MTBF value is not really a valid and useful technique for selecting task intervals.

The foregoing discussion has briefly outlined the most ideal situation that we experience for selecting task intervals. This ideal is not encountered as often as we would like to see because we usually do not have sufficient data from operating experience to define the fdf. So let's discuss what we can do in the non-ideal situations that are more commonly encountered. The first situation is one wherein we have a partial knowledge of the age–reliability relationship. This means that the failure cause information on the FMEA leads us to conclude that aging or wearout mechanisms are at play. Or perhaps we have some operating experience to support the conclusion that aging/wearout mechanisms exist. But, in either case, we do not have any statistical data to define when this would be expected to occur. So we tend to use our experience to guess at a task interval for the TD actions. In so doing, there is overwhelming evidence to show that this process is highly conservative. That is, we tend to pick intervals that are way too short. We might overhaul a large electric motor every three years when, in reality, the correct interval turns out to be 10 years! We must learn to correct this conservatism because it is costing us dearly. We do so via Age Exploration, which is described subsequently.

Again referring to Figure 5.21, the second situation is one in which we have no idea what the age–reliability relationship might be, and we are now moving on to look for candidate CD tasks. If the failure mode is hidden, we also extend our search to include candidate FF tasks. These tasks, too, must have intervals specified for the non-intrusive data acquisition and inspection actions that must be accomplished. And, here again, the statistical basis for specifying these intervals is usually missing, and we guess at what they will be—and usually with great conservatism. So Age Exploration will be useful to us with CD and FF tasks as well as with TD tasks. Let us offer one cautionary note about CD tasks. When we select a CD task, we must specify not only the task interval, but also the parameter value that must be used to alert the plant personnel that the incipient failure process has begun. Selecting the correct value may also be a guessing process at first, and additional experience must be systematically collected to adjust that value over time so that the alert is neither too early nor too late.

When good statistical data is not available, using our experience to guess at task intervals is really the only option that is available to us initially. But there is a proven technique that we can employ to refine that "guesstimate" over time, and to predict more accurately the correct task interval. It is called Age Exploration, or AE. The AE technique is strictly empirical, and works like this (using a TD task for illustrative purposes). Say our initial overhaul interval for a fan motor is 3 years. When we do the first overhaul, we meticulously inspect and record the as-found condition of the motor and all of its parts and assemblies where aging and wearout are thought to be possible. If our inspection reveals no such wearout or aging signs, when the next fan motor comes due for overhaul we automatically increase the interval by 10% (or more), and repeat the process, continuing until, on one of the overhauls, we see the incipient signs of wearout or aging. At this point, we stop the AE process, perhaps back off by 10%, and define this as our final task interval.

Figure 5.28 illustrates how this AE process was successfully used by United Airlines for one of their hydraulic pumps. On the top half of Figure 5.28, we see that the overhaul interval started at about 6000 hours, and that the AE process was then employed over a four-year period to extend the interval to 14,000 hours! The bottom half of Figure 5.28 presents a second very interesting statistic for the same population of pumps over the same four-year interval. The statistic is premature removal rate (or the rate at which corrective maintenance actions were required). The interesting point here is that the premature removal rate has a definite decreasing value over the four-year period where the overhaul interval was increasing. We interpret this to suggest that as the amount of human handling and intrusive overhaul maintenance actions decreased, so did the human error resulting from such actions, with the net effect that corrective maintenance actions likewise decreased. Recall that we saw this same human error effect in the statistics presented in Figure 2.2.

While it is certainly true that age exploration can be a lengthy process, one should consider that it is really the best alternative available when the statistical process cannot be used. As a rule, the collection of large samples of statistical data can be an even longer process.

5.10 ITEMS OF INTEREST (IOI)

As our early experience with the RCM process developed, we discovered a very pleasant surprise that was happening on virtually every systems analysis that was conducted. This surprise was the accrual of a series of peripheral technical and cost benefits that simply fell out from the rigors and thoroughness that are inherent to the RCM process. We captured these pearls of wisdom under the title "Items of interest," and recorded them as they occurred on an IOI form such as that shown on Figure 5.29. These "gratis" benefits were occasionally so significant that they alone were estimated to have paid for the cost of the systems analysis several times over. When people speak of the cost-benefits derived from an RCM program, these peripheral benefits are almost never included in their figures since future predictability of such benefits is elusive, and thus difficult to claim as expected credit. Nonetheless, it is the authors' view that such peripheral benefits will continue to occur, and thus represent just one more convincing argument why management should embrace the RCM methodology.

To drive this point home, several of the actual benefits that have been experienced are briefly discussed in order to illustrate the type of gratis fallout that you might

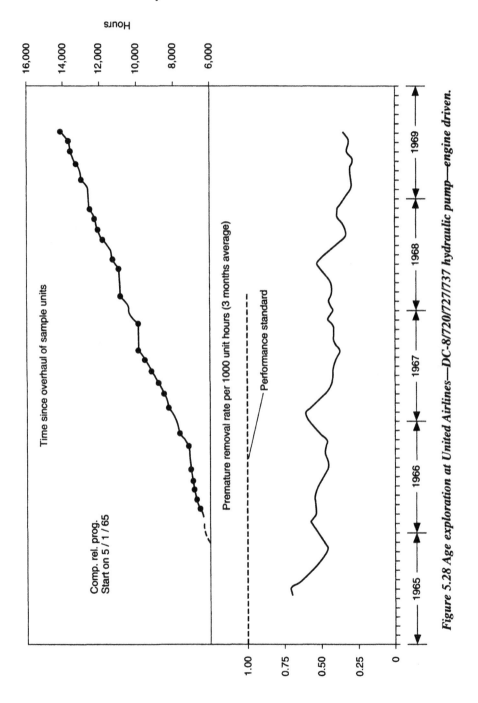

Figure 5.28 Age exploration at United Airlines—DC-8/720/727/737 hydraulic pump—engine driven.

RCM - Systems Analysis			
Information: Items Of Interest		**Rev No :**	
Plant:		**Date:**	
System:			
Subsystem:			
Analysts:			

IOI #	Date	Description	Status

System:	Tuesday, April 22, 2003
Subsystem:	Page 1 of 1
Items of Interest	
JMS Software	

Figure 5.29 Items of interest form.

see in your RCM program. We can put these benefits into one of five categories as a means of further identifying their favorable impact:

- Operational impact (OI)
- Safety impact (SI)
- Logistics impact (LI)
- Configuration impact (CI)
- Administration impact (AI)

All of them have a favorable cost impact, but this is not specified here (with one exception) because these items were not cost-quantified, as a rule, by the organizations involved. In retrospect, this was probably a mistake that should be corrected in future programs.

The list of examples is presented below, and their area of impact is indicated parenthetically in each case.

1. When completing the SWBS in Step 3 of the systems analysis process, the analyst discovered that the latest revision to the system drawings contained a change in the model number for one of the system's components. This led to an investigation to verify which model number should be used in the analysis. The final answer confirmed that the new model was correct—but it also revealed that several spares for the old model number had inadvertently been mistakenly held for some four years at an estimated cost of $75,000 to cover taxes, storage, and administrative expenses, before the RCM program triggered the actions necessary to dispose of the useless spare parts. (LI and AI.)

2. When the analyst was preparing the comparison of the RCM versus current PM tasks in Step 7-3, it was necessary to research the documentation defining the current PM actions. In doing so, it was discovered that the operations and maintenance departments were both performing some identical PM actions on the same component—but with different periodicity such that neither group ever discovered the duplication that was regularly occurring. The RCM process had resulted in a recommendation to do essentially the same task, so its implementation was instrumental in assigning the task responsibility solely to the maintenance department. (OI and AL.)

3. There have been instances where the analysts have discovered that a system contains one or more components that serve no useful role whatsoever in supporting the functions that have been identified for the system. Worse yet, this group of components frequently contains failure modes which can result in plant forced outages should they occur. The analysts tend to discover such situations at two points in the process: first, during the compilation of the SWBS in Step 3-4 where a detailed correlation is developed between the functional subsystems and the system P&ID; and, second, during the development of the equipment–functional failure matrix in Step 5-1, where it is first discovered that an empty set exists in the matrix. As a rule, these components are removed from the system at the earliest possible convenience, which is usually the first scheduled outage after support engineering has verified the analyst's finding. In many cases, the components involved in these problems are valves of one type or another (flow control, check, block, etc.) with solenoid valves, limit switches, instrumentation, and other peripheral items also involved. (OI, SI, LI, CI, AI.)

4. The rigors of the systems analysis process, at virtually every step along the way, require a detailed review of the system documentation. This, in turn, very often identifies either missing or erroneous information in the system baseline definition records. Such information includes corrections or additions to system P&IDs, system equipment and parts

lists, configuration control files, maintenance management information systems files, equipment tags, etc. Correction of such information is particularly significant in nuclear power plants or any facility where safety systems are key players in the everyday operations. (OI, SI, CI.)

5. The rigors of the process are also capable of identifying simple design enhancements that can eliminate failure modes, failure effects, and/or PM tasks. Consider the following examples: (1) upgrading an analog to a digital control device or instrument; (2) adding manual isolation valves to allow repair/replacement of failed items in redundant configurations (the authors never cease to be amazed at the frequency of such simple omissions in the original design); and (3) installation of the capability for manual addition of chemicals to automatic water-conditioning systems that fail. (OI, SI, CI.)

6. If the plant or system is new, the RCM process offers some excellent opportunities to correct deficiencies that were overlooked during the pre-operational punch-list walkdown and checkout. (And yes, these deficiencies do occur in the real world in spite of TQM, zero defects, TLC, and all of the other popular buzzwords that currently dominate our corporate management philosophy!) Two examples illustrate this point: (1) a system walkdown during Step 2-2 of the systems analysis process discovered that valves needed to activate an air-removal subsystem had not been opened (the immediate effect was inconsequential, but the long-term effect was potentially very damaging due to corrosion concerns); and (2) improper piping connections had made a water chemistry analyzer inoperative. Notification to operations corrected both situations in a timely fashion. (OI, CI.)

7. One RCM program used Step 7—task selection—as an opportunity also to further highlight where special instructions, cautions, and/or training requirements were needed to successfully implement the tasks. This addition to the task selection process is particularly noteworthy in light of the industrywide deficiencies associated with the lack of good documentation on standard maintenance practices and procedures. (SI, AI.)

8. Occasionally, Step 5-2—FMEA—will discover a heretofore unrecognized failure scenario that could initiate or directly produce serious operational or safety consequences. The likelihood of such a discovery during the RCM process is admittedly small, but the fact is that the FMEA presents an excellent opportunity to revisit a variety of operational conditions that in many older plants have not been reviewed for years. There have been recorded incidents of such discoveries in the systems analysis process and, in one case at a nuclear power plant, the discovery was made even after the completion of an extensive probabilistic risk assessment (PRA) on the system in question. (OI, SI.)

In addition to the preceding specific examples, virtually every organization that has conducted an RCM program has recognized the value of the systems analysis process as a training ground for system engineers. The training, in fact, is so comprehensive that the analysts often become the recognized "resident system expert" for those systems that have experienced the RCM process. Another related recognized value is the use of the systems analysis information in operator training exercises wherein, for example, failure scenarios developed in the FMEA would be used as simulator inputs to test operator response to plant transient or upset conditions. Both of these training benefits need to be more thoroughly developed by management in order to realize the full potential that they offer.

6

ILLUSTRATING RCM—A SIMPLE EXAMPLE (SWIMMING POOL MAINTENANCE)

The best way to illustrate any analysis process is by way of example. Thus, we will devote this chapter to just such an endeavor, and will illustrate the RCM systems analysis process with an application to a home swimming pool (in this case, the pool in Mac Smith's backyard at his prior residence of 24 years in Saratoga).

It is instructive to note that when Mac first acquired a home swimming pool, knowledge of how to maintain it was essentially zero. So he applied all of the current ad hoc PM methods (see Sec. 2.5) to formulate the program—that is, experience (of which he had none), judgment (of which he had much, being an engineer), vendors' and friends' recommendations (which he later learned were based on experience and judgment), and a bit of brute force (if the filter could be dismantled, that must be the correct thing to do frequently). That was in 1975. By 1985, Mac was beginning to see the light! His early dealings with RCM prompted him (slowly, but surely) to change the PM style. When Mac moved in 1999, he left behind a 28-year-old pool that still looked like new; a pool in which the water had never been removed and over 90 percent of the original components were intact and working like a charm.

In Chapter 6, we will follow the seven-step systems analysis process that was described in Chapter 5. But please recognize that some of the steps are rather easy to complete in comparison to what you would likely encounter in a more complex facility. Nonetheless, the principles are the same, and the illustration should help you to understand both the process and the mechanics of its implementation.

133

Also, note that the forms and information appearing in Steps 2 to 7 were generated using the "RCM WorkSaver" software (see Sec. 11.5).

6.1 STEP 1—SYSTEM SELECTION AND INFORMATION COLLECTION

6.1.1 System Selection

The typical home swimming pool can be conveniently viewed as consisting of four major systems:

1. The *pool system* proper is where the fun takes place.
2. The *spa system* adjoins or is adjacent to the pool system. Not all pools will have this particular feature, but in this case there is an adjoining or attached spa enclosure that is integral to the pool proper. That is where relaxation takes place.
3. The *water treatment system* can be most easily identified at this point in our discussion as the group of equipment usually hidden somewhere. That is where we keep the water "the way it should be."
4. The *utility system* supplies the electricity, gas, and water for the pool and its supporting equipment.

Steps 2, 3, and 4 will further describe these systems.

The system selection process in this example is quite easy because the only system with any significant equipment diversity (thus PM diversity) is the water treatment system. It can also be said from an 80/20 point of view that the CM costs are concentrated in the water treatment system. Thus selection criteria using a Pareto diagram would lead us to the water treatment system.

6.1.2 Information Collection

As far as available information is concerned, this was not a lengthy task to perform because there wasn't much documentation available in the first place. In fact, the only information passed to Mac as the second owner of the pool was a handwritten set of instructions on how to align the valves in order to heat and use the spa. The previous owner did conduct a walkdown which consisted of his collective O&M knowledge from five years of experience. The walkdown lasted all of 10 minutes, and basically emphasized the need to throw some muriatic acid into the pool every now and then, and all else would take care of itself. (As it turned out, this was the point to remember because, with hard water and rainfall as the water makeup source for evaporation, maintaining a neutral pH in the pool water is a key factor in avoiding any water problems.) Thus, for this illustration, it was necessary to re-create the system schematic and virtually all of the information that is needed to perform the RCM systems analysis.

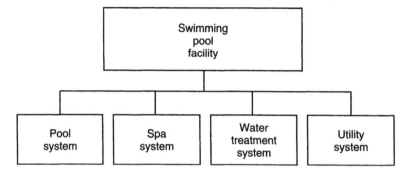

Figure 6.1 Swimming pool facility system block diagram.

At this point, it would not be surprising if some of you who are reading this book aren't sitting there with a little smile on your face. Why? Because this scenario doesn't sound all that different from what you have experienced with your plant or facility—especially if it has been there for some time. Generally, in an older facility or plant (say, 10 years or older), basic information on the system P&ID, system design descriptions, O&M manuals, and most OEM component manuals (even part lists) may not be readily available. Fortunately, in most situations, there are plant personnel on site who have the essential elements of this data stored either in their desks or in their minds. Also, OEM representatives stand ready to supply some information. Unfortunately, in this example, there was no prior experience still available to tap, and the OEM representatives came in the form of the local pool supply store, and in some instances, Mac's neighbors. Suffice it to say, as noted earlier, information collection may be one of the more difficult tasks to accomplish in the whole systems analysis process, especially when dealing with older facilities that are more complex than a swimming pool.

Two essential pieces of information that were re-created in order to guide both the system selection and Steps 2 and 3 of the process were the pool block diagram and a pool schematic shown in Figures 6.1 and 6.2, respectively. All remaining information was developed, as needed, for the subsequent steps in the process.

6.2 STEP 2—SYSTEM BOUNDARY DEFINITION

Boundary definition for the water treatment system will follow the format and content described in Chapter 5, Figures 5.5 and 5.6. The system schematic in Figure 6.2 was used to specify the system boundary. The resulting Boundary Overview and Boundary Details are given in Figures 6.3 and 6.4.

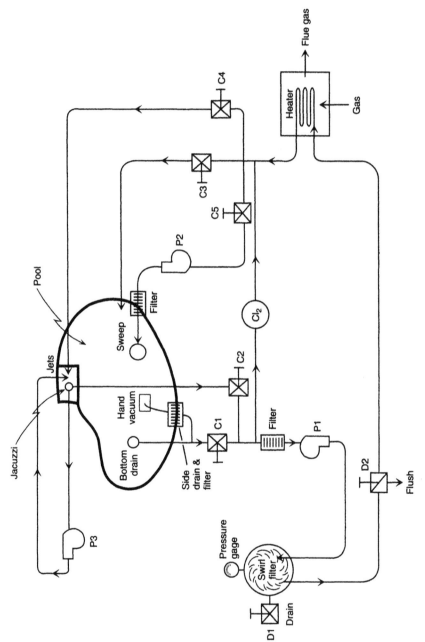

Figure 6.2 Swimming pool schematic.

RCM - Systems Analysis

Step 2-1:	**System Boundary Definition**	**Plant ID:**	
Information:	**Boundary Overview**	**System ID**	
Plant:	Swimming Pool Facility	**Rev No:**	
System:	Water Treatment	**Date:**	1/9/00
Subsystem:			
Analysts:	A. M. (Mac) Smith, G. R. Hinchcliffe		

Major Equipment Included:

Pumps and motors
Heat exchanger
Filters
Valves and piping
Chlorinator
Various instruments

Primary Physical Boundaries

Start with:

Water entrances to the pool and spa
Natural gas entrance to the heater
Electricity exiting the circuit breaker box

Terminate with:

Water exits from the pool and spa
Flue gas exit from the heater

Caveats:

None

System: Water Treatment	Wednesday, March 26, 2003
Subsystem:	Page 1 of 1

Step 2-1 Boundary Overview
JMS Software

Figure 6.3 Boundary overview, Step 2-1.

RCM - Systems Analysis

Step 2-2:	System Boundary Definition	Plant ID:	
Information:	Boundary Details	System ID	
Plant:	Swimming Pool Facility	Rev No:	
System:	Water Treatment	Date:	1/9/00
Subsystem:			
Analysts:	A. M. (Mac) Smith, G. R. Hinchcliffe		

Type	Bounding System	Interface Location	Reference Drawing
In (Pool Water)	Pool	Main water drain at bottom of pool	
In (Pool Water)	Pool	Water skimmer at side of pool	
Out (Filtered Water)	Pool	Return water inlets (two) at side of pool	
In (Spa Water)	Spa	Main water drain at bottom of spa	
Out (Heated Water)	Spa	Return water inlet at side of spa	
Out (Water Overflow)	Sewer/drainage	Exit water drain on main filter	
In (Gas)	Utility	Inlet side of gas valve on heater	
Out (Flue Gas)	Atmosphere	Exit gas ducts/openings on heat exchanger	
In (Electric)	Utility	Pool side of electrical circuit breaker on 120 V supply	
Out (Pool Water)	Pool	Quick disconnect on water line to the pool sweep	
Out (Flush Water)	Sewer/drainage	Exit water flush line on main filter	

System: Water Treatment
Subsystem:

Wednesday, March 26, 2003
Page 1 of 1

Step 2-2 Boundary Details
JMS Software

Figure 6.4 Boundary details, Step 2-2.

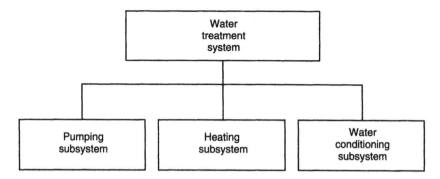

Figure 6.5 Functional block diagram for water treatment system.

6.3 STEP 3—SYSTEM DESCRIPTION AND FUNCTIONAL BLOCK DIAGRAM

6.3.1 Step 3-1—System Description

In this example, we can expand upon the pool system block diagram in Figure 6.1 to define the subsystem block diagram for the water treatment system as shown in Figure 6.5. This shows that we can conveniently divide the selected system into three functional subsystems—pumping, heating, and water conditioning. Since this is a fairly simple system, we will not divide the system description into three subsystems, but rather will address the water treatment system as one entity.

Swimming pool facility description. As an introduction to the water treatment system description, we will first discuss some features of the overall swimming facility. The facility, shown in Figure 6.6, is a 34,000-gallon kidney-shaped pool that is typical of residential pool installations. In northern California (our location), it can be used for about six months each year at a water temperature of 70°F or higher without any artificial heating. The latter, without an enclosure, can be prohibitively costly, and virtually no one will heat a pool for year-round use. Pools are usually located for afternoon exposure to the sun, and in summer months the water temperature usually peaks out at about 85°F without any artificial heating.

The pool system has three functional subsystems:

1. *Water fill.* This is simply the replacement of evaporated water with a garden hose (or, in some cases, with built-in water fill lines) to maintain a specified water level. Some pools are covered with a lightweight mylar blanket when not in use in order to minimize evaporation. The pool in our example here does not use any cover since it tends to be more of a nuisance than any real help if the pool is used frequently in warm weather.
2. *Manual water treatment.* This consists of netting large debris (such as leaves), vacuuming the pool as needed, and adding muriatic acid to the

Figure 6.6 Pool and spa system.

water for pH control and an oxidizer to retain chlorine activity. The water chemistry is also checked periodically for pH level and chlorine content—at least weekly in hot weather and monthly in cold weather to guide the need for chemical additions.

3. *Pool sweep.* As the name suggests, this is a water-driven device which sweeps about the pool with two subsurface water lines which operate in continuous motion, stirring the water to keep pool dirt in suspension for filtering. In recent years, some pools have installed automatic vacuum sweeps which continuously move about the pool sides and bottom with a suction action.

Many pools also have a spa system, and this is functionally, and often physically, tied to the main pool system. In this case, the spa system is a 6 × 3 × 3 foot rectangular enclosure with two walls in common with the main pool. The water is usually heated in the 90 to 120°F range, depending on personal preferences, and is a very enjoyable form of relaxation at the end of "one-of-those-days." The spa system has two functional subsystems:

1. *Manual clean.* This is the same as the preceding, for debris removal and vacuuming.
2. *Water jets.* Those of you familiar with spas know that high-velocity streams of water are pumped through "jets" in side locations to maintain a water circulation and swirl. In this spa system, there are three such jets. (The pressure behind these jets is not a part of the water treatment system.)

The utility system consists of the water, electricity, and gas services that supply the swimming pool facility.

Water treatment system (WTS). The water treatment system is responsible for the majority of the functions required to maintain water purification. It also provides for artificial heating of the pool and spa water. These functions are achieved via three functional subsystems—namely, pumping, heating, and water conditioning. Our description here, which follows the format of Figure 5.8, is shown in Figure 6.7.

RCM - Systems Analysis

Step 3-1:	**System Description/Functional Block Diagram**	**Plant ID:**
Information:	**System Description**	**System ID:**
Plant:	Swimming Pool Facility	**Rev No:**
System:	Water Treatment	**Date:** 1/9/00
Subsystem:		
Analysts:	A. M. (Mac) Smith, G. R. Hinchcliffe	

Functional Description/Key Parameters

PUMPING: The pumping subsystem provides two primary functions. First, it maintains a water flow at about 70 GPM circulating from the pool system through the heater and water conditioning subsystems. This flow includes a bleed through the line supplying water to the pool sweep in order to maintain a continuous priming flow for the pump in this line. Thus, the pool sweep cannot be operated unless the main water flow is in operation. Second, it provides the water flow and boost pressure, on demand, for the operation of the pool sweep. The pumping subsystem operates about 5 hours each day in the warm and hot seasons, and about 3 hours each day in the cooler months. These periods of operation are accomplished via automatic electromechanical switches that can turn both the main flow and pool sweep flow on and off at preset times. The pumping subsystem must be maintained in an airtight condition on the suction side of the water lines to preclude a loss of flow to the pumps. Also, during heavy rains in the winter months, the pumping subsystem must be able to drain water from the pool to avoid pool overflow.

HEATING: The heating subsystem provides the capability to raise the ambient temperature of either the pool or spa water. Water circulation from the pumping subsystem continuously flows through a heat exchanger which is operated on natural gas, and when ignited, has an output rating of 383,000 Btu/hour. The heat exchanger is automatically controlled to provide the desired temperature to the pool (about 80 deg F) or the spa (90 to 120 deg F). It cannot efficiently heat both the pool and spa simultaneously. Since, by choice, virtually no heating of the pool occurs, the pool temperature control is simply maintained at a setting that is well below the ambient water temperature. The control unit also has a "hi-limit: temperature switch which will stop operation if exceeded. This high limit is set at about 140 deg F. Ignition is via a gas pilot flame which is shut off during cooler months when neither the pool nor spa are used. The heating subsystem must be operated in a safe manner-that is, there must be no potential for personnel injury or death, or equipment damage from an uncontrolled fire or explosion.

System:	Water Treatment	Friday, April 04, 2003
Subsystem:		Page 1 of 2
	Step 3-1 System Description	
	JMS Software	

Figure 6.7 System description, Step 3-1.

WATER CONDITIONING: The water conditioning subsystem provides continuous automatic filtering and chlorination treatment. Its function, then, is to maintain the water in a crystal clear condition. The filtering is augmented by periodic manual netting and vacuuming of the pool system, and by periodic manual addition of muriatic acid and oxidizer to the pool system. But the mainstay of the water-conditioning process is the daily operation of the pumping subsystem, which maintains the flow through the filter equipment and the chlorinator. Coarse filtering occurs at two locations: first, through a filter basket at the weir/skimmer water exit at the side of the pool (most of the water exit flow occurs here, not at the bottom drain in the pool), and second, through a filter basket immediately ahead of the main pump suction. Fine filtering is accomplished via a 70-gallon capacity swirl filter which uses diatomaceous earth as a filter medium to remove dust and particles not stopped by the basket filters. The swirl filter (so called because the internal design includes a series of semicircular plastic sections that rotate the water flow across their surfaces which are coated with the diatomaceous earth) is located on the discharge side of the main pump, and contains both a drain valve for use in removing excess water from the pool and flush the swirl filter. A pressure gage on the swirl filter is used to calibrate the need for backflushing. The automatic chlorinator is simply a dispenser containing 1-inch chlorine tablets with a bleed line connected between the suction side piping and the heater exit piping. The bleed flow through this line can be adjusted up to 0.5 GPM to maintain a desired chlorine level in the pool water.

Redundancy Features

With the exception of the two in-line basket filters, the water treatment system has no redundancy features.

Protection features

There are two important protection features. First, any electrical malfunction of consequence in the motors or instrumentation will trip the circuit breaker, thereby preventing catastrophic damage or fire. Second, the high-limit temperature control will prevent inadvertent excessive heating of the pool or spa water (and unnecessary gas consumption) should the desired temperature set point malfunction. There is also a grace period of 3-7 days (depending on water temperature) for the maintenance of acceptable water quality should any failure cause a complete shutdown of the water treatment system. This is accomplished by manual additions of liquid chlorine and other fungus-retarding chemicals.

Key Control Features

The key instrumentation features of this subsystem are the electromechanical timers which provide for automatic operation of the pumping and water conditioning subsystems, and the high-limit temperature switch in the heating subsystem which was described previously. A manual switch is used to select heating for either the pool or spa, and a pressure gage is used to indicate flow status and clogging in the swirl filter.

System:	Water Treatment	Friday, April 04, 2003
Subsystem:		Page 2 of 2
	Step 3-1 System Description JMS Software	

Figure 6.7 Continued

6.3.2 Step 3-2—Functional Block Diagram

In Figure 6.5, we saw that the water treatment system had been divided into the three functional subsystems: pumping, heating, and water conditioning. In Figure 6.8, we now complete the functional block diagram by also including the IN and OUT interfaces as well as the crucial interconnecting interfaces.

6.3.3 Step 3-3—IN/OUT Interfaces

Using the format shown in Figure 5.10, we now list all of the appropriate IN/OUT interfaces on Figure 6.9. Keep in mind that the RCM process assumes that IN interfaces are available when needed, and therefore we will concentrate later on

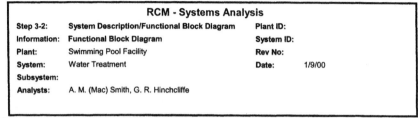

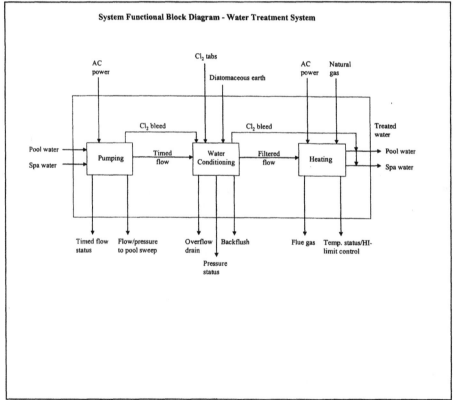

Figure 6.8 Functional block diagram, Step 3-2.

RCM - Systems Analysis

Step 3-3:	**System Description/Functional Block Diagram**	**Plant ID:**	
Information:	**IN/OUT Interfaces**	**System ID**	
Plant:	Swimming Pool Facility	**Rev No:**	
System:	Water Treatment	**Date:**	1/9/00
Subsystem:			
Analysts:	A. M. (Mac) Smith, G. R. Hinchcliffe		

Type	Bounding System	Interface Location	Reference Drawing
In (Pool Water)	Pool	Inlet side of Pool and Spa Botton Drain and Skimmer (up stream of valve C1)	
In (Spa Water)	Spa	Inlet side of Spa Drain and Skimmer (up stream of valve C2)	
In (AC power)	Utility	120 V from load side of pool system breaker to pumps and heater controls	
In (Natural Gas)	Utility	House main supply on line side of Pool Heater gas shut off valve	
In (Chlorine and Diatomaceous Earth)	Manual loading	Clorinator and Swirl Filter	
Out (Treated Water)	Pool	Outlet side of pool water return inputs (down stream of valve C3)	
Out (Treated Water)	Pool	Outlet side of Pool Sweep return inputs (down stream of valve C5, and pump P2)	
Out (Treated Water)	Spa	Outlet side of Spa jets (down stream of valve C4)	
Out (Signal Status)	Manual reading	Pumps, filter, and heater (.ie. timer, pressure, temperature)	
Out (Flue Gas)	Atmosphere	Exit gas duct from heater	
Out (Water)	Utility (city drainage)	Outlet of drain valve D1 on filter (upon demand)	
Out (Water)	Utility (sump pit)	Outlet of drain valve D2 on filter (upon demand)	

System: Water Treatment		Friday, April 04, 2003
Subsystem:		Page 1 of 2

Step 3-3 IN/OUT Interfaces
JMS Software

Figure 6.9 IN/OUT interfaces, Step 3-3.

using the OUT interfaces to identify and focus on function preservation and, ultimately, the selection of the PM tasks.

6.3.4. Step 3-4—System Work Breakdown Structure (SWBS)

In the SWBS, we list the specific components that are associated with each of the three functional subsystems. The SWBS for the water treatment system is shown in Figure 6.10.

RCM - Systems Analysis

Step 3-4:	System Description/Functional Block Diagram	Plant ID:	
Information:	System Work Breakdown Structure	System ID	
Plant:	Swimming Pool Facility	Rev No:	
System:	Water Treatment	Date:	1/9/00
Subsystem:			
Analysts:	A. M. (Mac) Smith, G. R. Hinchcliffe		

Comp	Component ID	Component Description	Type	Qty	Dwg or Ref
01		Main pump (P1) with 1-HP motor	Non-Instr	1	
02		Pool sweep pump (P2) with 3/4-HP motor	Non-Instr	1	
03		Valves-pool/spa alignment	Non-Instr	4	
04		Valve drain	Non-Instr	1	
05		Main Pump - Timer Electromechanical	C - Instr	1	
06		Pool Sweep - Timer Electromechanical	C - Instr	1	
07		Water piping	Non-Instr	various	
08		Main (swirl) filter	Non-Instr	1	
09		Trap filters	Non-Instr	2	
10		Chlorinator	Non-Instr	1	
11		Flush valve on main filter	Non-Instr	1	
12		Pressure gage on main filter	S - Instr	1	
13		Gas heater	Non-Instr	1	
14		Gas piping	Non-Instr	various	
15		Temperature control/limit unit	P - Instr		
16		Pool/spa switch	C - Instr	1	

| System: | Water Treatment | Friday, April 04, 2003 |
| Subsystem: | | Page 1 of 1 |

Step 3-4 SWBS
JMS Software

C=Control S=Status P=Protection

Figure 6.10 SWBS, Step 3-4.

6.3.5. Step 3-5—Equipment History

The objective here was to recall (there are no formal work orders) the corrective maintenance (CM) actions that have occurred in the water treatment system. We will use this data, as applicable, in Step 5 when constructing the failure mode information. The equipment history information, as reconstructed from repair bills and personal repair experiences, is shown in Figure 6.11. As a point of interest, notice that the water treatment system operated virtually free of any unexpected problems (i.e., corrective maintenance actions) during the first 5–7 years of operation during Mac's ownership (the equipment was 10–12 years old).

RCM - Systems Analysis

Step 3-5:	System Description/Functional Block Diagram	Plant ID:	
Information:	Equipment History	System ID	
Plant:	Swimming Pool Facility	Rev No:	
System:	Water Treatment	Date:	1/9/00
Subsystem:			
Analysts:	A. M. (Mac) Smith, G. R. Hinchcliffe		

Component Description	Failure Mode	Failure Cause
1. Alignment Valves	Stuck in closed/open position (1980, 88)	Corrosion on stem
2. Pinhole leaks in suction pumping	Connecting joint deterioration (1981, 87, 91)	Aging
3. Main pump	Bearing (sealed) breakdown (1988)	Aging
4. Main (swirl) filter-top canister	Lip fracture at joint with bottom canister (1986)	Material flaw (manufacturer replaced with no charge)
5. Main (swirl) filter-C clamp (top to bottom canister)	Overstressed (1982)	Human error- excessive tightening (Oops!)
6. Flush valve on main filter	Stuck closed (1984)	Lack of lubrication
7. Main filter pressure gage	Erratic reading (1984)	Seal leak to atmosphere
8. Gas heater	Erratic burner ignition (1989)	Contamination, corrosion

System:	Water Treatment	Friday, April 04, 2003
Subsystem:		Page 1 of 1

Step 3-5 Equip History
JMS Software

Figure 6.11 Equipment history, Step 3-5.

Even then, most of the corrective maintenance problems (items 1, 5, 6, 7, 8) could have been avoided with timely and correctly applied preventive maintenance!

6.4 STEP 4—SYSTEM FUNCTIONS AND FUNCTIONAL FAILURES

We will now use the information developed in the system descriptions, IN/OUT interfaces, and functional block diagram to formulate the specific function and functional failure statements. Much of our effort to this point has been directed toward the ability to accurately list functions and functional failures in order to properly guide the eventual selection of the PM tasks.

The information so structured is shown in Figure 6.12, which follows the format of Figure 5.13. Notice that we have started a numbering system as shown below which will be used in succeeding steps to maintain traceability.

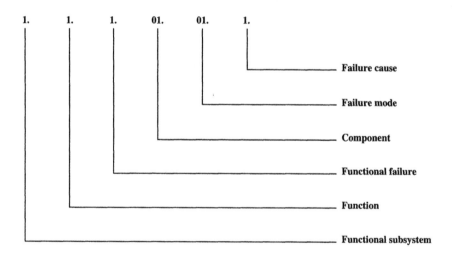

The first digit, which represents the functional subsystem, shows "1" for the pumping subsystem, "2" for the water conditioning subsystem, and "3" for the heating subsystem in Figure 6.12.

6.5 STEP 5—FAILURE MODE AND EFFECTS ANALYSIS

The initial action in Step 5 is to complete the equipment–functional failure matrix shown in Figure 5.14. We do this by combining the SWBS listings in

		RCM - Systems Analysis	
Step 4:	Functions/Functional Failures	Plant ID:	
Information:	Functional Failure Description	System ID	
Plant:	Swimming Pool Facility	Rev No:	
System:	Water Treatment	Date:	1/9/00
Subsystem:			
Analysts:	A. M. (Mac) Smith, G. R. Hinchcliffe		

Function #	FF #	Function/Functional Failure Description
1.1		*Maintain 70-GPM water flow at specified times to other subsystems*
	1.1.1	Fails to initiate flow at specified time
	1.1.2	Flow is less that 70 GPM
	1.1.3	Fails to terminate flow at specified time
1.2		*Maintain 50-GPM water flow at specified times to pool sweep line*
	1.2.1	Fails to initiate flow at specified time
	1.2.2	Flow is less that 50 GPM
	1.2.3	Fails to terminate before main flow shut-down
1.3		*Maintain water bleed to chlorinator*
	1.3.1	No water bleed flow
1.4		*Automatically activate/deactivate water flow*
	1.4.1	"On" and/or "Off" signals malfunction
2.1		*Provide filtered water to the heating subsystem*
	2.1.1	Fails to catch larger debris
	2.1.2	Poor filtering efficiency (can be related to FF no. 1.1.2)
2.2		*Send chlorinated water to exit piping*
	2.2.1	Fails to add chlorine to bleed water
	2.2.2	No bleed water flow
3.1		*Provide desired heat input to water, on demand, at 383,000 Btu/hour*
	3.1.1	Fails to ignite

System:	Water Treatment		Wednesday, March 26, 2003
Subsystem:			Page 1 of 2

Step 4 Functional Failures

JMS Software

Figure 6.12 Functions and functional failures, Step 4.

Function #	FF #	Function/Functional Failure Description
	3.1.2	Fails to shut down at desired temperature
3.2		*Maintain a safe operation*
	3.2.1	Uneven burn and gas accumulation
	3.2.2	Fails to shut down at Hi-limit control temperature
	3.2.3	Full/partial stoppage of flue gas release

System: Water Treatment		Wednesday, March 26, 2003
Subsystem:		Page 2 of 2
	Step 4 Functional Failures	
	JMS Software	

Figure 6.12 Continued

Figure 6.10 with the functional failure information in Figure 6.12 to produce the matrix shown in Figure 6.13. This matrix, then, becomes the road map to guide us in the FMEA, and is the "connecting tissue" between the functions and equipment. As the matrix shows, each item of equipment is associated with at least one functional failure, and 60 percent of the listed components are involved in two or more functional failures. Thus, it would be premature for us to judge that any of these components could be labeled *noncritical* and discarded from further consideration at this point in time (see Sec. 5.6 for some background discussion on this comment).

The FMEA is shown in Figure 6.14, and the information presented here is at the heart of the RCM process because it now identifies the specific failure modes that can potentially defeat our functions as delineated in Figure 6.12. There are 41 unique failure modes listed, and they lead to 37 cases for carryover to the logic tree analysis (LTA). Due to the lack of any significant redundancy in the water treatment system, only four cases are dropped at this point in the analysis (and two of these because the failure mode in question is considered implausible).

RCM - Systems Analysis

Step 5-1:	Functions/Functional Failure	Plant ID:
Information:	Equip - Functional Failure Matrix	System ID
Plant:	Swimming Pool Facility	Rev No:
System:	Water Treatment	Date: 1/12/00
Subsystem:		
Analysts:	A. M. (Mac) Smith, G. R. Hinchcliffe	

Functional Failures

Comp#	Component Description	1.1.1	1.1.2	1.1.3	1.2.1	1.2.2	1.2.3	1.3.1	1.4.1	2.1.1	2.1.2	2.2.1	2.2.2	3.1.1	3.1.2	3.2.1	3.2.2	3.2.3
01	Main pump (P1) with 1-HP motor	X	X															
02	Pool sweep pump (P2) with 3/4-HP m				X	X												
03	Valves-pool/spa alignment	X																
04	Valve drain	X																
05	Main Pump - Timer Electromechanic	1.4.1		1.4.1					X									
06	Pool Sweep - Timer Electromechanic				1.4.1		1.4.1		X									
07	Water piping	X			X	X		X					X					
08	Main (swirl) filter		2.1.2								X							
09	Trap filters		2.1.2							X	X							
10	Chlorinator											X						
11	Flush valve on main filter										X							
12	Pressure gage on main filter										X							
13	Gas heater													X		X		
14	Gas piping													X	X	X		
15	Temperature control/limit unit													X	X		X	
16	Pool/spa switch																	X

System:	Water Treatment	Step 5-1 Equip- FF Matrix	Wednesday, March 26, 2003
Subsystem:		JMS Software	Page 1 of 1

Figure 6.13 Equipment functional failure matrix, Step 5-1.

RCM - Systems Analysis

Step 5-2	Failure Mode and Effects Analysis	Plant ID:	
Information:	Functional Failure #	System ID:	
Plant:	Swimming Pool Facility	Rev No:	
System:	Water Treatment	Date:	1/12/00
Subsystem:			
Analysts:	A. M. (Mac) Smith, G. R. Hinchcliffe		

FF#	Comp	Component Description	FM #	Failure Mode	FC #	Failure Cause	Failure Effect Local	System	Plant	LTA
1.1.1	01	Main pump (P1) with 1-HP motor	01.01	Failed bearing (sealed)	01.01.1	Age/wearout	Pump inoperative	Loss of flow	Pool/spa water deterioration-spa inoperative	YES
1.1.1	01	Main pump (P1) with 1-HP motor	01.02	Motor short/ground	01.02.1	Insulation aging	Pump inoperative	Loss of flow	Pool/spa water deterioration-spa inoperative	YES
1.1.1	01	Main pump (P1) with 1-HP motor	01.03	Leak at pump motor joint	01.03.1	Broken gasket or loose bolts	Loss of pump suction- Possible motor/pump damage	Loss of flow	Pool/spa water deterioration-spa inoperative	YES
1.1.2	01	Main pump (P1) with 1-HP motor	01.04	Bearing deterioration	01.04.1	Age/ wearout	Erratic pump operation	Reduced flow	Pool/ spa water deterioration	YES
1.2.1	02	Pool sweep pump (P2) with 3/4-HP motor	02.01	Failed bearing (sealed)	02.01.1	Age/ wearout	Pump inoperative	Loss of flow to pool sweep	Pool/ spa water deterioration	YES
1.2.1	02	Pool sweep pump (P2) with 3/4-HP motor	02.02	Motor short/ ground	02.02.1	Insulation aging	Pump inoperative	Loss of flow to pool sweep	Pool/ spa water deterioration	YES
1.2.1	02	Pool sweep pump (P2) with 3/4-HP motor	02.03	Leak (at pump motor joint)	02.03.1	Broken basket or loose bolts	Loss of pump suction - possible pump/ motor damage	Loss of flow to pool sweep	Pool/ spa water deterioration	YES
1.2.2	02	Pool sweep pump (P2) with 3/4-HP motor	02.04	Bearing deterioration	02.04.1	Age/ wearout	Erratic pump operation	Reduced flow to pool sweep	Pool/ spa water deterioration	YES
1.1.1	03	Valves-pool/spa alignment	03.01	Stuck in open or closed position	03.01.1	Corrosion/ contamination	Cannot open or close valve on demand	cannot realign flow for pool to spa or vice versa	Either pool or spa inoperative	YES

System:	Water Treatment
Subsystem:	

Step 5-2 Failure Mode and Effects Analysis
JMS Software

Wednesday, March 26, 2003
Page 1 of 4

Figure 6.14 FMEA, Step 5-2.

FF#	Comp	Component Description	FM #	Failure Mode	FC #	Failure Cause	Failure Effect			
							Local	System	Plant	LTA
1.1.1	04	Valve drain	04.01	Stuck in closed (normal) position	04.01.1	Corrosion	Cannot open valve on demand	Cannot drain water from system	Pool/ spa can overflow in heavy rain	YES
1.4.1	05	Main Pump - Timer Electromechanical	05.01	Failed clock	05.01.1	Age/ wearout	Loss of automatic timing	System fails to start or fails to stop	Pool/ spa water deterioration for failed start. None for failed stop	YES
1.4.1	05	Main Pump - Timer Electromechanical	05.02	Short circuit	05.02.1	Insulation aging	Loss of automatic timing	System fails to start or fails to stop	Pool/ spa water deterioration for failed start. None for failed stop	YES
1.4.1	05	Electromechanical timers	05.03	Set points (mechanical) loose	05.03.1	Wear/ vibration or improperly installed	Loss of automatic timing	System fails to start or fails to stop	Pool/ spa water deterioration for failed start. None for failed stop	YES
1.4.1	06	Pool Sweep - Timer Electromechanical	06.01	Failed clock	06.01.1	Age/ wearout	Loss of automatic timing	Pool sweep fails to start or fails to stop and motor can burn out	Pool/ spa water deterioration	YES
1.4.1	06	Pool Sweep - Timer Electromechanical	06.02	Short circuit	06.02.1	Insulation aging	Loss of automatic timing. Note: Timer can be manually operated if someone observes failure state	Pool sweep fails to start or fails to stop and motor can burn out	Pool/ spa water deterioration	YES
1.4.1	06	Pool Sweep - Timer Electromechanical	06.03	Set points (mechanical) loose	06.03.1	Wear/ vibration or improperly installed	Loss of automatic timing. NOTE: Timer can be manually operated if someone observes failure state	System fails to start or fails to stop	Pool/ spa water deterioratio	YES
1.1.1	07	Water piping	07.01	Rupture	07.01.1	Material flaw-Considered to be implausible failure mode	Considered to be implausible failure mode	Considered to be implausible failure mode	Considered to be implausible failure mode	Rare Event

System: Water Treatment
Subsystem:

Step 5-2   Failure Mode and Effects Analysis
JMS Software

Wednesday, March 26, 2003
Page 2 of 4

Figure 6.14 Continued

FF#	Comp	Component Description	FM#	Failure Mode	FC#	Failure Cause	Local	Failure Effect System	Plant	LTA
1.1.1	07	Water piping	07.02	Pinhole leak (including joints)	07.02.1	Corrosion	Loss of pump suction (could temporarily restore)	Flow deterioration	Pool/spa water deterioration	YES
1.2.2	07	Water piping	07.03	Clogged filter (on pool sweep line)	07.03.1	Debris buildup	Reduced filter efficiency	Reduced flow to pool sweep/overworked pump	Pool/spa water deterioration	YES
1.3.1	07	Water piping	07.04	Rupture (3/8" neoprene bleed line)	07.04.1	Age	Loss of pump suction (could temporarily restore)	Flow deterioration	Pool/spa water deterioration	YES
1.3.1	07	Water piping	07.05	Pinhole leak - including joints (3/8" neoprene bleed line)	07.05.1	Age	Loss of pump suction (could temporarily restore)	Flow deterioration	Pool/spa water deterioration	YES
2.2.2	07	Water piping	07.06	Rupture (3/8" neoprene bleed line)	07.06.1	Age	Broken exit line from chlorinator	No chlorine injection to pool return piping	Pool/spa water deterioration	YES
2.1.2	08	Main (swirl) filter	08.01	Clogged	08.01.1	Debris and dirt buildup	Reduced filter efficiency	Reduced flow/overworked pump	Pool/spa water deterioration	YES
2.1.2	08	Main (swirl) filter	08.02	Water leak (at top to bottom section joint)	08.02.1	Aging gasket	Water dripping from filter joints	None	None	NO
2.1.1	09	Trap filters	09.01	Broken basket (at main pump suction)	09.01.1	Mishandle or age	Hole in plastic basket	Large debris escapes to swirl filter	None	YES
2.1.1	09	Trap filters	09.02	Broken basket (at weir)	09.02.1	Mishandle or age	Hole in plastic basket	Large debris escapes to second trap filter	None	YES
2.1.1	09	Trap filters	09.03	Clogged (at main pump suction)	09.03.1	Debris buildup	Reduced filter efficiency	Reduced flow/overworked pump	Pool/spa water deterioration	YES
2.1.1	09	Trap filters	09.04	Clogged (at weir)	09.04.1	Debris buildup	Reduced filter efficiency	Reduced flow/overworked pump	Pool/spa water deterioration	YES
2.1.1	09	Trap filters	09.05	Leaky gasket - on filter cover (at main pump suction)	09.05.1	Age	Loss of pump suction (could temporarily restore)	Flow deterioration	Pool/spa water deterioration	YES
2.2.1	10	Chlorinator	10.01	Clogged	10.01.1	Debris from undissolved tabs	No flow from chlorinator	No chlorine injection to bleed line	Pool/spa water deterioration	YES

Step 5-2   Failure Mode and Effects Analysis

JMS Software

System: Water Treatment

Subsystem:

Figure 6.14 Continued

FF#	Comp	Component Description	FM #	Failure Mode	FC #	Failure Cause	Local	Failure Effect		
								System	Plant	LTA
2.2.1	10	Chlorinator	10.02	No tablets	10.02.1	Forgot to fill	Empty chlorinator	No chlorine injection to bleed line	Pool/ spa water deterioration	YES
2.1.2	11	Flush valve on main filter	11.01	Stuck	11.01.1	Corrosion	Inoperative backflush valve	Cannot backflush main filter	None	YES
2.1.2	12	Pressure gage on main filter	12.01	False reading (lower than actual)	12.01.1	Age	Erroneous pressure signal	Reduced flow/ overworked pump. deterioration. Note: Assumes main filter is clogged	Pool/ spa water deterioration	YES
3.1.1	13	Gas heater	13.01	Failed pilot light	13.01.1	Wind or rain storm	No pilot light and smell of gas	Heater will not ignite on demand	Spa cannot be used	YES
3.2.1	13	Gas heater	13.02	Burner dirty/ clogged	13.02.1	Corrosion, dirt, insects	Delay in simultaneous ignition across burner	Small to large explosion in heater, possibly fire and severe heater damage	Spa cannot be used	YES
3.2.3	13	Gas heater	13.03	Clogged vents	13.03.1	Leaves, pine needles, insects, etc.	Blocked vents	Flue gas cannot escape - possible fire and damaged heater	Spa cannot be used	YES
3.1.1	14	Gas piping	14.01	Blockage (in heater piping)	14.01.1	Large foreign object in line	Considered to be implausible failure mode	Considered to be implausible failure mode	Considered to be implausible failure mode	Rare Event
3.2.1	14	Gas piping	14.02	Leak (at connection)	14.02.1	Age/ vibration	Smell of gas	None	None	NO
3.1.1	15	Temperature control/limit unit	15.01	Control unit fails (Lo)	15.01.1	Random part failure	Control unit inoperative	Heater will not ignite on demand	Spa cannot be used	YES
3.1.2	15	Temperature control/limit unit	15.02	Control unit fails (Hi)	15.02.1	Random part failure	Control unit inoperative	Heater will not automatically shut down	Eventual over temperature in spa; unnecessary gas consumption	YES
3.1.1	16	Pool/spa switch	16.01	Failed switch	16.01.1	Aging	No electrical contact through switch	Gas valve will not open - heater will not ignite	Spa cannot be used	YES

System: Water Treatment
Subsystem:

Step 5-2 Failure Mode and Effects Analysis
JMS Software

Friday, April 04, 2003
Page 4 of 4

Figure 6.14 Continued

6.6 STEP 6—LOGIC (DECISION) TREE ANALYSIS

We can use the LTA structure shown in Figure 5.20 for the LTA in this swimming pool illustration. In this case, the "operators" were Mac and his beloved better-half, and the *outage condition associated with question 3 in Figure 5.20 will be defined as the inability to use either the pool or the spa.* As we have seen in the FMEA in Figure 6.14, a majority of the "plant" effects are "pool/spa water deterioration" which, in their own right, are not the immediate initiators of outages. So we will see what happens to these failure modes as we progress through the LTA and then the sanity check in Step 7-2 of the systems analysis process.

The LTA information is shown in Figure 6.15, where 37 failure modes from the FMEA were carried over to the logic tree process. In summary, the LTA revealed the following categories (hence priorities):

A or D/A = 2
B or D/B = 8
C or D/C = 27

As suggested in Sec. 5.7, our initial action will be to relegate the 27 category C or D/C failure modes to the RTF status and a sanity check in Step 7-2, and to pass the 10 critical failure modes on to Step 7-1 task selection.

6.7 STEP 7—TASK SELECTION

6.7.1 Step 7-1—Task Selection Process

The category A, D/A, B, and D/B items from the LTA are put through the task selection process described in Figures 5.21 and 5.22. The results of this process are shown in Figure 6.16. Of the ten top-priority failure modes derived from the LTA, we defined two TD tasks, one CD task, two FF tasks and five RTF decisions. The latter RTF decisions were driven in three cases by the fact that no applicable task could be identified, and in two cases by the effectiveness consideration (see Sec. 4.4 for discussion of "Applicable and Effective").

6.7.2 Step 7-2—Sanity Check

Figure 6.17 lists each component and its related failure mode that was assigned as a category C or D/C priority from the LTA. The listing also contains the two items that were dropped at the FMEA (i.e., the main swirl filter section leak and the gas piping leak). Of the 29 failure modes subjected to the sanity check, 13 retained the RTF status while 16 items were assigned a PM task because of the marginal effectiveness or secondary damage potential associated with an

RCM - Systems Analysis

Step 6	Logic Tree Analysis	Plant ID:
Information:	Failure Mode Criticality	System ID
Plant:	Swimming Pool Facility	Rev No:
System:	Water Treatment	Date: 1/19/00
Subsystem:		
Analysts:	A. M. (Mac) Smith, G. R. Hinchcliffe	

FF #	Comp #	Component Description	FM #	Failure Mode	Evident?	Safety	Outage	Cat	Comments
1.1.1	01	Main pump (P1) with 1-HP motor	01.01	Failed bearing (sealed)	YES	NO	YES	B	Must be corrected in 4 or less days or serious water deterioration occurs
1.1.1	01	Main pump (P1) with 1-HP motor	01.02	Motor short/ground	YES	NO	YES	B	Must be corrected in 4 or less days or serious water deterioration occurs
1.1.1	01	Main pump (P1) with 1-HP motor	01.03	Leak at pump motor joint	YES	NO	YES	B	Must be corrected in 4 or less days or serious water deterioration occurs. Can cause serious motor/ pump damage if not shutoff in 4 or less hours
1.1.2	01	Main pump (P1) with 1-HP motor	01.04	Bearing deterioration	YES	NO	NO	C	Gives audible indication
1.2.1	02	Pool sweep pump (P2) with 3/4-HP motor	02.01	Failed bearing (sealed)	YES	NO	NO	C	
1.2.1	02	Pool sweep pump (P2) with 3/4-HP motor	02.02	Motor short/ ground	YES	NO	NO	C	
1.2.1	02	Pool sweep pump (P2) with 3/4-HP motor	02.03	Leak (at pump motor joint)	YES	NO	NO	C	Can cause serious motor/ pump damage if not shutoff in 4 or less hours
1.2.2	02	Pool sweep pump (P2) with 3/4-HP motor	02.04	Bearing deterioration	YES	NO	NO	C	Gives audible indication
1.1.1	03	Valves–pool/spa alignment	03.01	Stuck in open or closed position	NO	NO	YES	D/B	
1.1.1	04	Valve drain	04.01	Stuck in closed (normal) position	NO	NO	YES	D/B	Spa cannot be used if pool water level is too high

System:	Water Treatment	Wednesday, March 26, 2003
Subsystem:		Page 1 of 3

Step 6   **Logic Tree Analysis**
JMS Software

Figure 6.15 Logic tree analysis, Step 6.

FF #	Comp #	Component Description	FM #	Failure Mode	Evident?	Safety	Outage	Cat	Comments
1.4.1	05	Main Pump - Timer Electromechanical	05.01	Failed clock	YES	NO	NO	C	Main concern is failure to start. Could lead to pool sweep motor damage when pool sweep timer initiates
1.4.1	05	Main Pump - Timer Electromechanical	05.02	Short circuit	YES	NO	NO	C	Main concern is failure to start. Could lead to pool sweep motor damage when pool sweep timer initiates
1.4.1	05	Electromechanical timers	05.03	Set points (mechanical) loose	YES	NO	NO	C	Main concern is failure to start. Could lead to pool sweep motor damage when pool sweep timer initiates
1.4.1	06	Pool Sweep - Timer Electromechanical	06.01	Failed clock	YES	NO	NO	C	Main concern is failure to start. Could lead to pool sweep motor damage when pool sweep timer initiates
1.4.1	06	Pool Sweep - Timer Electromechanical	06.02	Short circuit	YES	NO	NO	C	Main concern is failure to start. Could lead to pool sweep motor damage when pool sweep timer initiates
1.4.1	06	Pool Sweep - Timer Electromechanical	06.03	Set points (mechanical) loose	YES	NO	NO	C	Main concern is failure to start. Could lead to pool sweep motor damage when pool sweep timer initiates
1.1.1	07	Water piping	07.02	Pinhole leak (including joints)	YES	NO	NO	C	Must be corrected in 4 or less days or serious water deterioration occurs. Can cause serious motor/ pump damage if not shutoff in 4 or less hours
1.2.2	07	Water piping	07.03	Clogged filter (on pool sweep line)	NO	NO	NO	D/C	Can shorten motor life if clogged for several months
1.3.1	07	Water piping	07.04	Rupture (3/8" neoprene bleed line)	YES	NO	NO	C	Must be corrected in 4 or less days or serious water deterioration occurs. Can cause serious motor/ pump damage if not shutoff in 4 or less hours
1.3.1	07	Water piping	07.05	Pinhole leak - including joints (3/8" neoprene bleed line)	YES	NO	NO	C	Must be corrected in 4 or less days or serious water deterioration occurs. Can cause serious motor/ pump damage if not shutoff in 4 or less hours
2.2.2	07	Water piping	07.06	Rupture (3/8" neoprene bleed line)	NO	NO	NO	D/C	Must be corrected in 4 or less days or serious water deterioration occurs
2.1.2	08	Main (swirl) filter	08.01	Clogged	YES	NO	NO	C	Pressure gage reading is increasing. Can shorten motor life if clogged for several weeks
2.1.1	09	Trap filters	09.01	Broken basket (at main pump suction)	NO	NO	NO	D/C	Increases debris buildup in main swirl filter

System:
Subsystem: Water Treatment

Step 6   Logic Tree Analysis
JMS Software

Wednesday, March 26, 2003
Page 2 of 3

Figure 6.15 Continued

FF #	Comp #	Component Description	FM #	Failure Mode	Evident?	Safety	Outage	Cat	Comments
2.1.1	09	Trap filters	09.02	Broken basket (at weir)	NO	NO	NO	D/C	Increases debris buildup in second trap filter
2.1.1	09	Trap filters	09.03	Clogged (at main pump suction)	NO	NO	NO	D/C	Could shorten motor life if clogged for several weeks
2.1.1	09	Trap filters	09.04	Clogged (at weir)	NO	NO	NO	D/C	Could shorten motor life if clogged for several weeks
2.1.1	09	Trap filters	09.05	Leaky gasket - on filter cover (at main pump suction)	YES	NO	NO	C	Must be corrected in 4 or less days or serious water deterioration occurs. Can cause serious motor/pump damage if not shutoff in 4 or less hours.
2.2.1	10	Chlorinator	10.01	Clogged	NO	NO	NO	D/C	Must be corrected in 4 or less days or serious water deterioration occurs
2.2.1	10	Chlorinator	10.02	No tablets	NO	NO	NO	C	Must be corrected in 4 or less days or serious water deterioration occurs
2.1.2	11	Flush valve on main filter	11.01	Stuck	NO	NO	NO	D/C	
2.1.2	12	Pressure gage on main filter	12.01	False reading (lower than actual)	NO	NO	NO	D/C	Would remove capability to easily discern that filter clogging is occurring
3.1.1	13	Gas heater	13.01	Failed pilot light	YES	NO	YES	B	Can smell gas
3.2.1	13	Gas heater	13.02	Burner dirty/ clogged	NO	YES	YES	D/A/B	
3.2.3	13	Gas heater	13.03	Clogged vents	YES	YES	YES	A/B	
3.1.1	15	Temperature control/limit unit	15.01	Control unit fails (Lo)	NO	NO	YES	D/B	
3.1.2	15	Temperature control/limit unit	15.02	Control unit fails (Hi)	YES	NO	NO	C	
3.1.1	16	Pool/spa switch	16.01	Failed switch	NO	NO	YES	D/B	

System: Water Treatment
Subsystem:

Step 6   Logic Tree Analysis
JMS Software

Wednesday, March 26, 2003
Page 3 of 3

Figure 6.15 Continued

RCM - Systems Analysis

Step 7-1	Task Selection		Plant ID:	
Information:	Selection Process and Decision		System ID	
Plant:	Swimming Pool Facility		Rev No:	
System:	Water Treatment		Date:	1/19/00
Subsystem:				
Analysts:	A. M. (Mac) Smith, G. R. Hinchcliffe			

FF #	Comp	Comp Desc	FM #	Failure Mode	FC #	Failure Cause	1	2	3	4	5	6	7	Candidate Task	Effective Info	Selective Dec.	Est Freq
1.1.1	01	Main pump (P1) with 1-HP motor	01.01	Failed bearing (sealed)	01.01.1	Age/wearout	N	N	Y	N	-	N	N	1. Vibration monitoring 2. RTF	1. Is not considered cost effective	RTF	-
1.1.1	01	Main pump (P1) with 1-HP motor	01.02	Motor short/ground	01.02.1	Insulation aging	N	N	N	N	-	N	N	1. RTF	There is no practical way to stop or detect onset of this failure mode	RTF	-
1.1.1	01	Main pump (P1) with 1-HP motor	01.03	Leak at pump motor joint	01.03.1	Broken gasket or loose bolts	N	N	N	N	-	Y	-	1. Inspect for signs of water seepage at the gasket area. (CD) 2. RTF	1. This is the most cost effective task	Inspect for signs of water seepage at the gasket area. (CD)	6 mo.
1.1.1	03	Valves-pool/spa alignment	03.01	Stuck in open or closed position	03.01.1	Corrosion/ contamination	P	Y	N	Y	Y	Y	-	1. Check the valve operation in early spring (before use of spa). (FF) 2. Lubricate valve stem. (TD) 3. RTF	1. Is the most cost effective since history indicates infrequent sticking	Check the valve operation in early spring (before use of spa). (FF)	12 mo. (spring)
1.1.1	04	Valve drain	04.01	Stuck in closed (normal) position	04.01.1	Corrosion	P	Y	N	Y	Y	Y	-	1. Check valve operation in early fall (before rainy season). (FF) 2. Exercise valve periodically and lubricate. (TD) 3. RTF	1. Is the most cost effective since history indicates no sticking 3. Did not choose RTF since sticking valve could lead to backyard flooding	Check valve operation in early fall (before rainy season). (FF)	12 mo. (fall)
3.1.1	13	Gas heater	13.01	Failed pilot light	13.01.1	Wind or rain storm	Y	N	N	N	-	N	N	1. RTF	There is no way to stop or detect onset of failed pilot light. Smell of gas near heater is very evident.	RTF	-

| System: | Water Treatment | Step 7-1   Task Selection | Wednesday, March 26, 2003 |
| Subsystem: | | JMS Software | Page 1 of 2 |

Figure 6.16 Task selection, Step 7-1.

FF #	Comp	Comp Desc	FM #	Failure Mode	FC #	Failure Cause	1	2	3	4	5	6	7	Candidate Task	Effective Info	Selective Dec.	Est Freq
3.2.1	13	Gas heater	13.02	Burner dirty/ clogged	13.02.1	Corrosion, dirt, insects	Y	Y	Y	N	-	Y	-	1. Remove burner unit and clean - repair as required. (TD) 2. Observe delay in ignition of all burners. (CD) 3. RTF	1. Appears to be the most cost effective. 2. It is questionable if this option is realistic. 3. RTF is not an option here due to category A rating.	Remove burner unit and clean - repair as required. (TD)	60 mo.
3.2.3	13	Gas heater	13.03	Clogged vents	13.03.1	Leaves, pine needles, insects, etc.	Y	Y	N	N	-	Y	-	1. Clean gas heater vents in early spring (before use of spa). (TD) 2. RTF	1. Is the most effective 2. RTF is not an option here due to category A rating.	Clean gas heater vents in early spring (before use of spa). (TD)	12 mo. (spring)
3.1.1	15	Temperature control/limit unit	15.01	Control unit fails (Lo)	15.01.1	Random part failure	N	N	N	N	N	N	N	1. RTF There is no applicable PM task for the failure cause.	1. Is the only option. Unit has not failed to date.	RTF	-
3.1.1	16	Pool/spa switch	16.01	Failed switch	16.01.1	Aging	N	N	N	Y	Y	N	N	1. Periodically check switch for proper operation. 2. RTF	2. Is most cost effective. Switch has never failed. If it did, replacement is quick and cheap.	RTF	-

System:	Water Treatment	Step 7-1 Task Selection	Wednesday, March 26, 2003
Subsystem:		JMS Software	Page 2 of 2

Figure 6.16 Continued

RCM - Systems Analysis

Step 7-2:	Task Selection
Information:	Sanity Checklist
Plant:	Swimming Pool Facility
System:	Water Treatment
Subsystem:	
Analysts:	A. M. (Mac) Smith, G. R. Hinchcliffe

Plant ID:	
System ID:	
Rev No:	
Date:	1/24/00

FF #	Comp #	Comp Desc	FM #	Failure Mode	Marg Eff	High Cost	Sec Dmg	OEM Conf	Int Conf	Reg Conf	Insur Conf	Hidden	RTF	Selection Dec	Est Freq	Comments
1.1.2	01	Main pump (P1) with 1-HP motor	01.04	Bearing deterioration	X							NO	NO	Listen for discernable increase in pump/motor noise level. (CD)	6 mo.	This was successfully used to predict need for motor/pump replacement in 1984 before total failure
1.2.1	02	Pool sweep pump (P2) with 3/4-HP motor	02.01	Failed bearing (sealed)								NO	YES	RTF	-	While some form of vibration monitoring may be possible, this is not considered to be a cost effective approach. Motor and pump are easily replaceable if required, but expected life is 15 to 20 years.

System:	Water Treatment
Subsystem:	

Step 7-2 Sanity Checklist
JMS Software

Wednesday, March 26, 2003
Page 1 of 4

Figure 6.17 Sanity checklist, Step 7-2.

FF #	Comp #	Comp Desc	FM #	Failure Mode	Marg Eff	High Cost	Sec Dmg	OEM Conf	Int Conf	Reg Conf	Insur Conf	Hidden	RTF	Selection Dec	Est Freq	Comments
1.2.1	02	Pool sweep pump (P2) with 3/4-HP motor	02.02	Motor short/ ground								NO	YES	RTF	-	A motor short, due to rain penetration or long-term deterioration, is impossible to prevent via any PM action, and onset is virtually impossible to detect.
1.2.1	02	Pool sweep pump (P2) with 3/4-HP motor	02.03	Leak (at pump motor joint)			X					NO	NO	Inspect for signs of water seepage at the gasket area. (CD)	6 mo.	
1.2.2	02	Pool sweep pump (P2) with 3/4-HP motor .	02.04	Bearing deterioration	X							NO	NO	Listen for discernible increase in pump/motor noise level. (CD)	6 mo.	
1.4.1	05	Main Pump - Timer Electromechanical	05.01	Failed clock								NO	YES	RTF	-	No applicable task.
1.4.1	05	Main Pump - Timer Electromechanical	05.02	Short circuit								NO	YES	RTF	-	No applicable task.
1.4.1	05	Electromechanical timers	05.03	Set points (mechanical) loose	X		X					NO	NO	Check to assure that set point screws are tight. (FF)	3 mo.	
1.4.1	06	Pool Sweep - Timer Electromechanical	06.01	Failed clock								NO	YES	RTF	-	No applicable task.
1.4.1	06	Pool Sweep - Timer Electromechanical	06.02	Short circuit								NO	YES	RTF		No applicable task.
1.4.1	06	Pool Sweep - Timer Electromechanical	06.03	Set points (mechanical) loose	X		X					NO	NO	Check to assure that set point screws are tight. (FF)	3 mo.	
1.1.1	07	Water piping	07.02	Pinhole leak (including joints)								NO	YES	RTF	-	No applicable task.

System: Water Treatment
Subsystem:

Step 7-2   Sanity Checklist
JMS Software

Wednesday, March 26, 2003
Page 2 of 4

Figure 6.17 Continued

FF #	Comp #	Comp Desc	FM #	Failure Mode	Marg Eff	High Cost	Sec Dmg	OEM Conf	Int Conf	Reg Conf	Insur Conf	Hidden	RTF	Selection Dec	Est Freq	Comments
1.2.2	07	Water piping	07.03	Clogged filter (on pool sweep line)	X							YES	NO	Remove and clean filter. (TD)	12 mo.	
1.3.1	07	Water piping	07.04	Rupture (3/8" neoprene bleed line)								NO	YES	RTF	-	No applicable task.
1.3.1	07	Water piping	07.05	Pinhole leak - including joints (3/8" neoprene bleed line)								NO	YES	RTF	-	No applicable task.
2.2.2	07	Water piping	07.06	Rupture (3/8" neoprene bleed line)								YES	YES	RTF	-	No applicable task.
2.1.2	08	Main (swirl) filter	08.01	Clogged	X							NO	NO	Monitor filter pressure gage reading for pressure increase above approximately 20 psi. (CD)	6 mo.	
2.1.2	08	Main (swirl) filter	08.02	Water leak (at top to bottom section joint)								NO	YES	RTF	-	No system or pool effect.
2.1.1	09	Trap filters	09.01	Broken basket (at main pump suction)	X							YES	NO	Remove filter basket, inspect and clean. Replace basket if necessary. (TD)	3 mo.	
2.1.1	09	Trap filters	09.02	Broken basket (at weir)	X							YES	NO	Remove filter basket, inspect and clean. Replace basket if necessary. (TD)	3 mo.	
2.1.1	09	Trap filters	09.03	Clogged (at main pump suction)	X							YES	NO	Remove filter basket, inspect and clean. (TD)	3 mo.	
2.1.1	09	Trap filters	09.04	Clogged (at weir)	X							YES	NO	Remove filter basket, inspect and clean. (TD)	3 mo.	
2.1.1	09	Trap filters	09.05	Leaky gasket - on filter cover (at main pump suction)	X							NO	NO	Replace gasket. (TD)	24 mo.	

System: Water Treatment
Subsystem:

Step 7-2 Sanity Checklist
JMS Software

Monday, April 28, 2003
Page 3 of 4

Figure 6.17 Continued

FF #	Comp #	Comp Desc	FM #	Failure Mode	Marg Eff	High Cost	Sec Dmg	OEM Conf	Int Conf	Reg Conf	Insur Conf	Hidden	RTF	Selection Dec	Est Freq	Comments
2.1.2	12	Pressure gage on main filter	12.01	False reading (lower than actual)	X							YES	NO	When monitoring gage for pressure reading, assure that gage returns to zero when main pump is turned off. (CD)	6 mo.	This visual check does not absolutely guarantee that the pressure gage is reading correctly, but observing its performance for erratic behavior when the pump is turned on and off is a reasonable indicator of unreliable readings. Pressure gage should be replaced if performance is suspect.
3.2.1	14	Gas piping	14.02	Leak (at connection)								NO	YES	RTF	-	No applicable task.
3.1.2	15	Temperature control/limit unit	15.02	Control unit fails (Hi)								NO	YES	RTF	-	No system or pool effect.
2.2.1	10	Chlorinator	10.01	Clogged	X							YES	NO	Remove and clean dispenser unit. (TD)	12 mo.	
2.2.1	10	Chlorinator	10.02	No tablets	X							NO	NO	Refill Cl2 tabs. (TD)	2 mo. (off season) 2 wk. (during season)	
2.1.2	11	Flush valve on main filter	11.01	Stuck	X							YES	NO	Operate valve and lubricate if necessary. (TD)	6 mo.	

System: Water Treatment
Subsystem:

Step 7-2   Sanity Checklist
JMS Software

Saturday, April 12, 2003
Page 4 of 4

Figure 6.17 Continued

RTF decision. In selecting the PM tasks here, the task selection process rationale in Figure 5.21 was employed, but the formality of completing the form was not accomplished. This form is a matter of choice, and could be used if the analyst so elected.

6.7.3 Step 7-3—Task Comparison

Figure 6.18 presents a summary of the system analysis results for each component in the water treatment system. The components are listed in the same order as in Figure 6.13, and the results are shown by failure mode per component. The two right-hand columns list the corresponding current task description for each component failure mode. The point that stands out here is that the current PM tasks were essentially representative of a reactive program. The large number of "none" entries (for 32 of the final list of 39 failure modes) says that maintenance was done, for the most part, by fixing things when they broke (i.e., corrective maintenance). In the current PM program, there were no deliberate decisions to "RTF" although the "none" entries could be considered as "RTF by default."

The RCM process introduced a rational PM program to the swimming pool and, quite frankly, helped to avoid a rash of bothersome (and sometimes harmful to the equipment) failure events. One example recalls going on a one-week vacation only to return and find that the main pump was not shutting off; the setscrew on the "off" switch came loose. It was impossible to know how long it had been running, but we were lucky that it was not the "on" switch that came loose because this would have damaged (maybe burned out) the pool sweep motor! It was also very pleasing to stop the biannual ritual of disassembling the main (swirl) filter for cleaning.

We can summarize the task comparisons as follows:

	By component			By failure mode	
	RCM	**Current**		**RCM**	**Current**
TD	5	4	TD	11	7
CD	4	0	CD	6	0
FF	4	0	FF	4	0
NONE	N/A	12	NONE	N/A	32
RTF	3	N/A	RTF	18	N/A

and

RCM = current (including components that are both RTF and none)   3
RCM = current, but modified   4

RCM - Systems Analysis

Step 7-3:	Task Selection	Plant ID:	
Information:	Comparison RCM vs Current PM Task	System ID	
Plant:	Swimming Pool Facility	Rev No:	
System:	Water Treatment	Date:	1/24/00
Subsystem:			
Analysts:	A. M. (Mac) Smith, G. R. Hinchcliffe		

FF#	Comp #	Comp Description	FM #	Fail Mode/Where From	RCM Selection Dec/Cat	Est Freq	Current Task Description	Freq
1.1.1	01	Main pump (P1) with 1-HP motor	01.01	Failed bearing (sealed) 7-1	RTF	-	None	-
1.1.1	01	Main pump (P1) with 1-HP motor	01.02	Motor short/ground 7-1	RTF B	-	None	-
1.1.1	01	Main pump (P1) with 1-HP motor	01.03	Leak at pump motor joint 7-1	Inspect for signs of water seepage at the gasket area. (CD) B	6 mo.	None	-
1.1.2	01	Main pump (P1) with 1-HP motor	01.04	Bearing deterioration 7-2	Listen for discernible increase in pump/motor noise level. (CD) C	6 mo.	None (The noise level detection was accidentally discovered in 1984)	-
1.2.1	02	Pool sweep pump (P2) with 3/4-HP motor	02.01	Failed bearing (sealed) 7-2	RTF	-	None	-
1.2.1	02	Pool sweep pump (P2) with 3/4-HP motor	02.02	Motor short/ ground 7-2	RTF C	-	None	-
1.2.1	02	Pool sweep pump (P2) with 3/4-HP motor	02.03	Leak (at pump motor joint) 7-2	Inspect for signs of water seepage at the gasket area. (CD) C	6 mo.	None	-
1.2.2	02	Pool sweep pump (P2) with 3/4-HP motor	02.04	Bearing deterioration 7-2	Listen for discernible increase in pump/motor noise level. (CD) C	6 mo.	None	-

Step 7-3 Comparison RCM vs Current PM Task
JMS Software

System:	Water Treatment
Subsystem:	

Wednesday, March 26, 2003
Page 1 of 4

Figure 6.18 Task comparison, Step 7-3.

FF#	Comp #	Comp Description	FM #	Fail Mode/Where From	RCM Selection Dec/Cat	Est Freq	Current Task Description	Freq
1.1.1	03	Valves-pool/spa alignment	03.01	Stuck in open or closed position / 7-1	Check the valve operation in early spring (before use of spa). (FF) / D/B	12 mo. (spring)	None (actually stuck twice, could not use spa until time was available to fix)	-
1.1.1	04	Valve drain	04.01	Stuck in closed (normal) position / 7-1	Check valve operation in early fall (before rainy season). (FF) / D/B	12 mo. (fall)	None	-
1.4.1	05	Main Pump - Timer Electromechanical	05.01	Failed clock / 7-2	RTF / C	-	None	-
1.4.1	05	Main Pump - Timer Electromechanical	05.02	Short circuit / 7-2	RTF / C	-	None	-
1.4.1	05	Electromechanical timers	05.03	Set points (mechanical) loose / 7-1	Check to assure that set point screws are tight. (FF) / C	3 mo.	None	-
1.4.1	06	Pool Sweep - Timer Electromechanical	06.01	Failed clock / 7-2	RTF / C	-	None	-
1.4.1	06	Pool Sweep - Timer Electromechanical	06.02	Short circuit / 7-2	RTF / C	-	None	-
1.4.1	06	Pool Sweep - Timer Electromechanical	06.03	Set points (mechanical) loose / 7-2	Check to assure that set point screws are tight. (FF) / C	3 mo.	None	-
1.1.1	07	Water piping	07.01	Rupture / Rare Event		-		
1.1.1	07	Water piping	07.02	Pinhole leak (including joints) / 7-2	RTF / C	-	None	-
1.2.2	07	Water piping	07.03	Clogged filter (on pool sweep line) / 7-2	Remove and clean filter. (TD) / D/C	12 mo.	None	-
1.3.1	07	Water piping	07.04	Rupture (3/8" neoprene bleed line) / 7-2	RTF / C	-	None	-
1.3.1	07	Water piping	07.05	Pinhole leak - including joints (3/8" neoprene bleed line) / 7-2	RTF / C	-	None	-

System: Water Treatment
Subsystem:

Step 7-3   Comparison RCM vs Current PM Task
JMS Software

Figure 6.18 Continued

FF#	Comp #	Comp Description	FM #	Fail Mode/Where From	RCM Selection	Dec/Cat	Est Freq	Current Task Description	Freq
2.2.2	07	Water piping	07.06	Rupture (3/8" neoprene bleed line)	RTF	D/C 7-2	-	None	-
2.1.2	08	Main (swirl) filter	08.01	Clogged	Monitor filter pressure gage reading (for pressure increase above approximately 20 psi. (CD)	C 7-2	6 mo.	Disassemble filter and clean. (TD)	6 mo.
2.1.2	08	Main (swirl) filter	08.02	Water leak (at top to bottom section joint)	RTF	C 7-2	-	None	-
2.1.1	09	Trap filters	09.01	Broken basket (at main pump suction)	Remove filter basket, inspect and clean. Replace basket if necessary. (TD)	D/C 7-2	3 mo.	Check filter baskets and clean after a storm. (TD)	Varied
2.1.1	09	Trap filters	09.02	Broken basket (at weir)	Remove filter basket, inspect and clean. Replace basket if necessary. (TD)	D/C 7-2	3 mo.	Check filter baskets and clean after a storm. (TD)	Varied
2.1.1	09	Trap filters	09.03	Clogged (at main pump suction)	Remove filter basket, inspect and clean. (TD)	D/C 7-1	3 mo.	Check filter baskets and clean after a storm. (TD)	Varied
2.1.1	09	Trap filters	09.04	Clogged (at weir)	Remove filter basket, inspect and clean. (TD)	D/C 7-2	3 mo.	Check filter baskets and clean after a storm. (TD)	Varied
2.1.1	09	Trap filters	09.05	Leaky gasket - on filter cover (at main pump suction)	Replace gasket. (TD)	C 7-2	24 mo.	None	-
2.2.1	10	Chlorinator	10.01	Clogged	Remove and clean dispenser unit. (TD)	D/C 7-2	12 mo.	None	-
2.2.1	10	Chlorinator	10.02	No tablets	Refill Cl2 tabs. (TD)	C 7-1	2 mo. (off season) 2 wk. (during season)	Refill Cl2 tabs. (TD)	2 mo. (off season) 2 wk. (during season)
2.1.2	11	Flush valve on main filter	11.01	Stuck	Operate valve and lubricate if necessary. (TD)	D/C 7-2	6 mo.	None	-

System: Water Treatment
Subsystem:

Step 7-3   Comparison RCM vs Current PM Task
JMS Software

Figure 6.18 Continued

FF#	Comp #	Comp Description	FM #	Fail Mode/Where From	RCM Selection	Dec/Cat	Est Freq	Current Task Description	Freq
2.1.2	12	Pressure gage on main filter	12.01	False reading (lower than actual) 7-2	When monitoring gage for pressure reading, assure that gage returns to zero when main pump is turned off. (CD)	D/C	6 mo.	None	-
3.1.1	13	Gas heater	13.01	Failed pilot light 7-1	RTF	B	-	None	-
3.2.1	13	Gas heater	13.02	Burner dirty/ clogged 7-1	Remove burner unit and clean - repair as required. (TD)	D/A/B	60 mo.	None	-
3.2.3	13	Gas heater	13.03	Clogged vents 7-1	Clean gas heater vents in early spring (before use of spa). (TD)	A/B	12 mo. (spring)	Clean gas heater vents in early spring (before use of spa). (TD)	12 mo. (spring)
3.1.1	14	Gas piping	14.01	Blockage (in heater piping) Rare Event					
3.2.1	14	Gas piping	14.02	Leak (at connection) 7-2	RTF	C	-	None	-
3.1.1	15	Temperature control/limit unit	15.01	Control unit fails (Lo) 7-1	RTF	D/B	-	None	-
3.1.2	15	Temperature control/limit unit	15.02	Control unit fails (Hi) 7-2	RTF	C	-	None	-
3.1.1	16	Pool/spa switch	16.01	Failed switch 7-1	RTF	D/B	-	None	-

System: Water Treatment
Subsystem:

Step 7-3 Comparison RCM vs Current PM Task
JMS Software

Wednesday, March 26, 2003
Page 4 of 4

Figure 6.18 Continued

RCM, but no current	9
Current, but no RCM	0

In other words, the RCM process directed more deliberate PM decisions where such had never occurred. The net result was trouble-free (virtually no corrective maintenance) operation for over seven years.

7

ALTERNATIVE ANALYSIS
METHODS

We have previously emphasized the need to focus PM resources where the ROI potential is greatest—i.e., on the 80/20 systems—and to employ the Classical RCM process to assure that the PM task structure on those systems will optimally impact reductions in corrective maintenance and plant downtime (see, for example, Secs. 1.4 and 5.1). This, however, leaves us with the question of what, if anything, could be done for the 20/80 systems. It also leaves unanswered the question of how one might reduce the costs of the Classical RCM process if resource constraints so dictated such consideration. In recent years, many forms of "shortcut" RCM analysis methods have emerged to answer these questions. The authors, together with several of our clients, have reviewed many of these shortcuts and have developed some concern about the effectiveness of the results that they often produce. This chapter discusses these issues, and provides specific solutions that are embodied in the Abbreviated Classical RCM™ process and the Experience-Centered Maintenance™ (ECM) process, both of which are described in detail.

7.1 REDUCING ANALYSIS COST

Everybody wants something for nothing. It seems that maintenance management looks upon the RCM process with much the same view. The hue and cry to reduce the cost of RCM analysis was so persistent that the Electric Power Research Institute (EPRI) embarked upon a project in the mid-1990s with the objective "to define specific methods for reducing the costs of performing RCM on nuclear plant systems." Information gathered at an EPRI utility workshop reduced the list of possible approaches to three specific methods: Streamlined Classical RCM Process, Plant Maintenance Optimizer Streamlined Process, and Critical Checklist Streamlined Process (the latter bearing little resemblance to RCM). These three "streamlined" methods were then evaluated at two nuclear power plants. The final reports from this project in 1995 made claims that the RCM analysis cost could be reduced by a factor of two to four in comparison to the Classical RCM process while producing high quality results (see Refs. 35 and 36).

The EPRI claims were based on a comparison of analysis costs for the three streamlined methods with average costs for prior Classical RCM analyses on other systems that were both complex and large. The three streamlined methods produced essentially identical results among themselves for the systems under evaluation in the project, but resource constraints precluded their direct comparison with a Classical RCM analysis on these same systems. Since then, there has been no EPRI project for direct comparison between streamlined and classical methods, and thus no hard evidence is available to support beyond reasonable doubt the claim made in Refs. 35 and 36.

Since the EPRI reports were published in 1995, the authors have observed several RCM "shortcut" methods where it was clear that the quality of results was poor and often of little value to the client. Others have made similar observations, such as John Moubray in his widely read article in *Maintenance Technology* magazine (Ref. 37). One characteristic that seems to be common to these shortcut methods is the very significant replacement of plant personnel in the analysis effort by outside contracted personnel on the RCM team. To the contrary, we believe that technical accuracy and plant buy-in with the analysis results can only be achieved via direct involvement of plant personnel at all stages of the analysis process (see Secs. 9.1, 9.2, and 9.3 for more thoughts on this). Also, when one considers the use of an RCM shortcut method, care must not only be exercised to require direct participation of plant personnel on the RCM team, but must likewise insure that certain crucial elements of the RCM process are present, especially if the system in question is an 80/20 system or one with considerable impact on plant performance and safety issues. These elements should include the following:

1. Does the method contain all four basic features that constitute RCM?
2. Does the method define a way to directly link function/functional failure with equipment failure modes?
3. Does the method include a direct means for identifying hidden failure modes?
4. Does the method provide a clear process for assigning criticality levels to each failure mode?
5. Does the method incorporate sufficient safeguards to assure that RTF decisions have been thoroughly evaluated?

MILL GRINDING (PULVERIZER) SUBSYSTEM

Systems Analysis Profile

	NEAL 3	CB 3
Number of Functions	6	2
Number of Functional Failures	14	2
Number of Components in the System Boundary	13	3
Number of Failure Modes Analyzed	130	8
- Hidden Failure Modes	88	0
Number of Critical Failure Modes	73	5
Number of PM Tasks Specified (incl RTF)	141	8
Number of "Items of Interest" (IOI's)	49	0

Figure 7.1 Mill grinding (pulverizer) subsystem—systems analysis profile.

In the absence of any known project that deliberately set out to objectively evaluate Classical versus shortcut RCM methods, we have searched for any realistic situation that might provide such an evaluation.

In 1996, quite by accident, such an objective test took place. MidAmerican Energy conducted a Classical RCM analysis on the coal pulverizer at its Neal 3 plant while a sister plant at Council Bluffs conducted a streamlined RCM analysis on the exact same pulverizer. The results of this comparison are dramatic (Ref. 38). An overview comparison of the two projects is shown in Figure 7.1, where the significant results of the analyses are summarized in a side-by-side tabulation. What is particularly striking here is that the streamlined RCM analysis cost was 80 percent of the cost realized on the Classical RCM analysis and did not approach the cost savings previously claimed. What is equally striking is that the Neal 3 maintenance (PM & CM) labor was reduced by 1333 hours annually and annualized MW-hours lost were reduced by 94 percent as measured by 3-year pre-versus-post RCM periods. In contrast, Council Bluffs had little, if any, adjustments made in their pulverizer PM program while Neal 3 had adjustments in 69 percent of their pre-RCM PM program (see Figure 7.2). Likewise, there was little, if any, corresponding positive effect on CM or MW hours lost in the pre-versus post streamlined RCM periods at Council Bluffs. We believe that these results speak for themselves in terms of our general concern over the use of shortcut RCM methodologies.

However, that still leaves unanswered the question of reducing Classical RCM analysis costs. We have taken two positive steps to address this issue. First, we formed a cooperative alliance with JMS Software of San Jose, CA, to develop the "RCM WorkSaver" software (see Chapter 11). This software was specially

MILL GRINDING SUBSYSTEM

PM TASK SIMILARITY PROFILE – NEAL 3
(FOR FAILURE MODES)

Similarity Descriptor	Number	√	Percent
1. RCM = Current (Tasks are identical)	9		6%
2. RCM = Modified Current	42	√	30%
3A. RCM Specified Task – NO Current Task Exists (Current missed important failure modes)	49	√	35%
3B. RCM Specified RTF – NO Current Task Exists (Similarity probably accidental in most cases)	35		25%
4. RCM Specified RTF – Current Task Exists (Current approach not cost effective)	2	√	1%
5A. Current Task Exists – No Failure Mode in RCM Analysis	0		0%
5B. Current Task Exists – RCM Specifies Entirely Different Task	4	√	3%
	141		100%

√ = % of PM tasks changed by RCM   √ = 69%

Figure 7.2 Mill grinding subsystem—PM task similarity profile.

tailored to meet a specification that would fulfill the Classical RCM methodology described in Chapter 5. Use of this software reduces the applied labor hours by about 20 percent because of the efficiencies it incorporates in areas such as the automatic rollover and recording of information that is repeated in the Classical RCM documentation process. Second, we developed an Abbreviated Classical RCM™ process which adheres to the major features of the full Classical process and also uses the same "RCM WorkSaver" software. The Abbreviated Classical RCM™ process reduces the applied labor hours by another 20 percent. The details provided in Sec. 7.2 describe exactly how this reduction is achieved.

But let's be careful about one very important point. We believe that the 80/20 systems will always deserve the full Classical RCM process since, by definition, the potential ROI here is very large and thus deserves this level of detailed attention. In the 20/80 systems (i.e., the more well-behaved systems where the ROI potential becomes diminishingly small), we would suggest the use of the Abbreviated Classical RCM™ process. In fact, where systems are very well behaved, and the only objective is possibly to provide minor adjustments to the existing PMs, we recommend use of the Experience-Centered Maintenance™ (ECM) process described in Sec. 7.3. ECM is clearly not RCM—but it is a tool for having a quick look at these very well-behaved systems.

7.2 ABBREVIATED CLASSICAL RCM™ PROCESS

Our basic intention with the Abbreviated Process is to retain the four RCM principles, and to also retain a lesser but sufficient degree of the Classical 7-step Process defined in Chapter 5 to assure a rather thorough review and analysis of the system and its equipment. We still want to define the "right" PM tasks to perform, and to adjust the PM program as may be needed. But as a well-behaved system, there is just so much blood that can be squeezed out of the turnip—so we don't want to expend too much of our resources to find it. After the pilot project has been completed, the reductions that have been defined for the Abbreviated Process will reduce the analysis time by about 20 percent vis-à-vis the Classical Process. This means that an Abbreviated Process that is done using the "RCM WorkSaver" software can be completed by the team in about 3 weeks including training, whereas the Classical Process is about 4 to 5 weeks long with its associated training. In some cases, the learning curve effect has reduced these times even further. However, the team always has the discretion to perform and document any of these deletions if it feels the time and effort to do so is warranted. Before proceeding, be sure that you are familiar with the 7-step process found in Chapter 5 in order to understand why each step can be modified as described below for the more well-behaved 20/80 systems.

7.2.1 Step 1—System Selection and Information Collection

The data analysis that has been performed as Step 1 in the Classical Process provides the information required to define the 80/20 and 20/80 systems (i.e. the Pareto diagrams). So the only effort needed now is to identify and locate the necessary documentation on our selected 20/80 systems that will be needed by the team to perform the analysis.

7.2.2 Step 2—System Boundary Definition

We still need to know what is or is not part of our selected system, but, in Step 2, the only documentation that we will provide is the Boundary Overview, using the form in Step 2-1. The Boundary Details in <u>Step 2-2 will not be formally addressed</u> or documented. This, of course, places a responsibility on the team members to mentally keep the boundary points in mind as they proceed with the remaining steps.

7.2.3 System Description and Functional Block Diagram

Step 3-1—System description. While the team will obviously need to discuss various aspects of the selected system, <u>none of this will be documented</u>. However, the team still needs to be sensitive to the need for an understanding of the salient points of the system—that is, Functional Description/Key Parameters, Redundancy Features, Protection Features, and Key Control Features.

Step 3-2—Functional block diagram. Much of our understanding of how we might need to divide a system into subsystems, how to develop functional interfaces with adjacent subsystems and systems, and how to visualize the IN/OUT interfaces is graphically displayed in the Functional Block Diagram. The Abbreviated Process <u>does retain the Functional Block Diagram</u> since this is a fairly easy and quick way for the team to summarize their view of the system, and their understanding of its functional role in the facility.

Step 3-3—IN/OUT interfaces. We certainly cannot overstate the important role that this step plays in the analysis since an accurate knowledge of the OUT interfaces directly leads the team to the system functions. However, in the Abbreviated Process, we will <u>not separately record these interfaces</u>. Rather, the team will rely on the Functional Block Diagram to portray this information.

Step 3-4—System work breakdown structure (SWBS). We must know what components are considered to be within the boundaries of our system, and will be analyzed in succeeding steps. So the team will take the information in Step 2-1 plus a working knowledge of the system, and <u>will completely document</u> Step 3-4 using the appropriate forms.

Step 3-5—Equipment history. We will <u>not formally address or document</u> any equipment history. As noted previously in Chapter 5, even in the Classical Process we rarely find sufficient and useful equipment history to warrant the use of Step 3-5.

7.2.4 Step 4—Functions and Functional Failures

We <u>will address and document</u> Step 4. The team will use the Functional Block Diagram from Step 3-2 to guide this part of the analysis. However, the team will need to carefully consider and discuss the information needed to accurately list all system functions since the Abbreviated Process has not formally considered Steps 2-2, 3-1, and 3-3, all of which are important preludes to Step 4 in the Classical Process.

7.2.5 Step 5—Failure Mode and Effects Analysis

Step 5-1—Functional failure–equipment matrix. This matrix is a key analysis tool in that it shows the specific relationship between potential component problems and their ability to cause one or more of the functional failures. We <u>will address and document</u> the matrix.

Step 5-2—Failure mode and effects analysis (FMEA). In terms of the detail that is required to properly make PM task decisions, the FMEA is an absolutely crucial and necessary piece of information. Thus, we <u>will address and document</u> the FMEA to the same extent and completeness that is done in the Classical Process.

7.2.6 Step 6—Logic (Decision) Tree Analysis (LTA)

Experience with the Classical Process has clearly demonstrated that the team's familiarity with the use of the LTA develops quickly. Within a few line items on the FMEA, the team can state the category assignment without the need to formally answer the three yes/no questions. So that is what we will do in the Abbreviated Process—namely, go directly on the form for Step 6 to the Category column and make the appropriate entry. (Note: When using Version 2.0 or higher of the "RCM WorkSaver" software, the YES/NO columns are automatically recorded to match the selected category.)

7.2.7 Step 7—Task Selection

Step 7-1—Task Selection. The Task Selection roadmap used in the Classical Process should still be employed by the team members to structure their thought process in choosing and documenting Candidate Applicable Tasks in the Abbreviated Process. But the yes/no answers to the roadmap are not recorded on the form. Likewise, the Effectiveness column on the form is almost always used in the Classical Process to give some indication of the reasoning behind the specific selection that is made from the Candidate Applicable Tasks. In the Abbreviated Process, we will not include any documentation in the Effectiveness column. The selection decision and estimated frequency are documented.

Step 7-2—Sanity check. We will address and document the so-called insignificant failure modes in exactly the manner that was done in the Classical Process.

Step 7-3—Task comparison. In the Abbreviated Process, we will not perform any Task Comparisons.

7.2.8 Items of Interest (IOI)

In the Abbreviated Process, we will address and document the IOIs in exactly the manner that was done in the Classical Process.

A summary comparison of the Classical and Abbreviated Classical RCM™ processes is shown in Figure 7.3.

7.3 EXPERIENCE-CENTERED MAINTENANCE™ (ECM) PROCESS

In dealing with the 20/80 (well-behaved) systems, we may decide that some of them are still sufficiently critical or important to our operation that we need to take a good look at them by using the Abbreviated Classical RCM™ process. But, as we move down the "loss chain," the need for some form of review diminishes considerably—perhaps to the point that no further review is warranted (if it ain't broke, don't fix it). However, one last form of maintenance analysis (ECM)

Comparison

RCM Steps	Classical	Abbreviated
• 1 System Selection	• YES	• YES
• 2 System Boundary	• YES	• MODIFIED
- 2.1 Boundary Overview	- YES	- YES
- 2.2 Boundary Details	- YES	- NO
• 3 System Description	• YES	• MODIFIED
- 3.1 System Description	- YES	- NO
- 3.2 Functional Block Diagram	- YES	- YES
- 3.3 In / Out Interfaces	- YES	- NO
- 3.4 Equipment List	- YES	- YES
- 3.5 System History	- YES	- NO
• 4 Functions/Functional Failures	• YES	• YES
• 5 FMEA	• YES	• YES
- 5.1 Comparison Matrix	- YES	- YES
- 5.2 FMEA	- YES	- YES
• 6 Decision Tree Analysis (LTA)	• YES	• YES
• 7 Task Selection	• YES	• MODIFIED
- 7.1 Task Selection	- YES	- REDUCED
- 7.2 Sanity Check	- YES	- YES
- 7.3 Task Comparison	- YES	- NO

Figure 7.3 Comparison between classical and abbreviated classical RCM processes.

is suggested here for those situations where management may feel the need to have a "quick look" at certain plant systems.

The ECM process is intended to spend a minimal amount of analysis effort to determine if what we currently do via PM tasks is worth it, and then whether we have possibly overlooked any reasonable PM opportunity to gain some cost-effective benefit. Three separate, but short, forms of analysis are employed. You will notice, as we describe these three analyses, that we have departed from the basic RCM principles for the most part, only to retain some of the underlying thought process, especially the selection of Applicable and Effective PM tasks. Each part of the analysis is triggered by a key question. And while no software package exists for this method, a simple spreadsheet could be developed and employed to contain the results of the ECM analysis.

7.3.1 Part A

Are the current PM tasks performed on this system (if any) really worth it (i.e., is each one Applicable and Effective)? We will develop a simple spreadsheet of data to answer this question, see Figure 7.4:

Column 1:   List the current PM tasks one at a time.

1	2	3	4	5	6	7
Current PM Tasks (1 per row)	Component Name	Specific Failure Mode Addressed by PM	If Applicable Describe Effect of Failure	Is PM Effective? (Y/N)	Keep Task? RTF? Drop Task? Modify Task?	Describe New or Change to PM & Interval
Inspect Cleanliness of motor air ducts	C92 Compressor Motor	Clogged air filter	Motor Overheating	Y	Modify Task	New PM to periodically replace filters at 1M w/AE
Simulate warning & trip levels	Lube Oil Low/High Temp. Switch	Digital alarm board failure	Loss of protection	Y	Keep	Keep task from MP-SD-00651-041402

Figure 7.4 Example ECM method "A".

Column 2:   What component(s) in the system are affected by this PM task?

Column 3:   Are there specific failure modes that are addressed by this PM task? If so, list them and be as specific in describing them as possible.

If we cannot identify any plausible failure mode (including what we call a rare event) that relates to this PM task, then this current task does not pass the Applicability test, and should be dropped. In other words, resources are being spent with no benefit realized.

Column 4:   If Column #3 above passed the Applicability test, then describe in a few words the effect of this failure mode.

Column 5:   In the team's judgment, does the potential loss (i.e., CM + DT) from the Effect in Column #4 outweigh the cost of the PM task (recognizing that the PM cost may have to be viewed on a multi-year basis to get a realistic comparison)? This judgment is essentially one of putting the Applicable PM task to the test of Effectiveness—is it cost effective to spend the PM dollars? Enter a brief statement on the reasoning and conclusion.

Column 6:   Here is where we put our final answer for the data accumulated in Columns #1 through #5. Three answers are possible:

YES—the PM task appears to be both Applicable and Effective. Keep the task.

RTF—while the PM task is Applicable, it is not Effective. Drop the task.

NO—the PM task is not even Applicable. Drop the task.

Column 7:   If Column 6 is "Yes," is there any task modification that could be recommended to improve it (e.g. increase the interval via Age Exploration)?

7.3.2 Part B

Could any of the corrective maintenance events of the past five years (use more or less years at the team's discretion) been avoided if a proper (Applicable) PM task had been in place? Again, a spreadsheet of data should be employed to answer this question, see Figure 7.5:

Column 1:   List the CM event date.

Column 2:   Specific component involved initially in the failure scenario (in some instances, multiple components may have been involved).

Column 3:   Briefly describe the CM event (attach additional description sheets if necessary).

Column 4:   Identify the specific component failure mode in the event.

1	2	3	4	5	6	7	8
CM Event Date	Component Name	CM Event Description	Failure Mode	Failure Cause	Failure Effect Description	Failure Effect Warrants PM (Y/N)	Describe New or Change to PM & Interval
9-Nov-89	C92 Compressor	Water leak	Air cooler gasket leak	Pinched	Loss of compressor	N	Installation error, PM task not feasible.
5-May-93	C92 Compressor	Water leak	Air cooler gasket minor leak	Vibration	Minor water spill	Y	Change PM XXX to include 'Verify water jacket bolt tightness"

Figure 7.5 Example ECM method "B".

Column 5:   Identify the specific failure cause for the failure mode in Column #4 (if possible).

Column 6:   Describe briefly the effect of this failure scenario.

Column 7:   Is the effect in Column #6 sufficiently severe enough to warrant some preventive action?—YES or NO.

Column 8:   If Column #7 is YES, define, if possible, an Applicable and Effective PM task and interval that should be introduced. Or, if one exists, how should it be changed to assure that it is Applicable? If NO, briefly explain.

7.3.3 Part C

Can the team hypothesize any failure modes, not already covered in Part A or Part B above, that could potentially produce severe consequences (i.e., Category A—Safety, or Category B—Outage consequences)? The following spreadsheet of data should be employed, see Figure 7.6:

Column 1:   List each component.

Column 2:   List each specific failure mode that has been hypothesized.

Column 3:   Identify the team's best estimate of the failure cause in Column #2.

Column 4:   Describe briefly the effect of this failure scenario.

Column 5:   Define, if possible, an Applicable and Effective PM task and interval that should be introduced.

1	2	3	4	5
Component Name	New Failure Mode	Failure Cause	Failure Effect Description	Describe New PM & Interval
Lube Oil High Temperature Switch	Meter drift	Age or vibration	Inaccurate indication, possible false trips	Calibrate - 6 M w/AE

Figure 7.6 Example ECM method "C".

A typical ECM analysis should take about 2 to 4 days using a team of craft personnel who are familiar with the equipment and operation of the system. Remember that the system is already known to be well behaved, so the likelihood of any significant changes to its current PM activity is small. But the ECM process could provide some fine tuning that results in a beneficial Return-On-Investment (ROI).

8

IMPLEMENTATION—CARRYING RCM TO THE FLOOR

Upon completion of the 7-step systems analysis process described in Chapter 5, we must now take the final and crucial action to realize the fruits of our efforts. We must carry the recommended RCM tasks to the floor. That is, we must implement and accomplish the Task Packaging that was first introduced in Chapter 2 and Figure 2.1.

The Task Packaging effort has all too frequently proven to be a difficult task to accomplish successfully and efficiently. In truth, Task Packaging is somewhat analogous to the dangers that are inherent to an iceberg—the clearly visible part is only the top 1/7th of the iceberg (i.e., the 7-step RCM process just completed), but extending far below the surface is the other 6/7ths of the iceberg (i.e., Task Packaging). Many RCM programs have gone afoul on this hidden hazard. And why is that? What was not consciously known and thus not properly addressed, that could create such difficulties?

Typically, an RCM program will initially focus its planning and execution on completing the systems analysis process—a very logical thing to do. But therein lies a major factor in the trap that we inadvertently set for ourselves. We have failed to realize the extent and complexity of the planning and coordination that must be achieved in order to implement the analysis results. When reality sets in (and it will—big time), the resulting hiccup causes delayed decisions, stalling and a variety of miscommunications that were not foreseen.

The authors' experience with successful RCM programs shows a very strong correlation between "success" and the amount of early-on planning for the actions required to carry the entire effort all the way to implementation on the floor. Avoiding the icebergs is achieved with diligent navigation (planning, support, and execution).

Utilizing our experience, we will chart a detailed course for avoiding the known hazards to arrive safely on the shore of a successfully implemented RCM program. We will demystify the steps that, if considered and executed early on, assure success, highlighting the problems that have traditionally plagued RCM programs, and suggest how to avoid or at least mitigate their effects. Having now charted a course that can steer clear of hidden hazards, we will guide you through the basics of having your new and improved PM program communicate effectively with your CMMS. Ending our discussion on RCM implementation, we will suggest one way of developing useful task procedures from the information contained in the RCM analysis.

Before we begin to chart our course through the ice fields, this would be a good time to remind the reader that RCM execution is not a one-time, once-through event. Instead, RCM is a philosophy and a journey—*RCM is a paradigm shift in how maintenance is perceived and executed*. No matter at what stage you find your PM program—the beginning with a pilot project to see if RCM is for you, in the middle implementing the improvements on your critical and not-so-critical systems, or later on down the road keeping vigil and executing your living RCM program (see Chapter 10)—a truly cost-effective PM program can be realized by executing the concepts outlined in this chapter.

8.1 HISTORICAL PROBLEMS AND HURDLES

When Step 7 of the systems analysis process has been completed, the PM tasks that will optimize the plant PM program have been specified—i.e., those tasks which will produce the best ROI. The rigor of the PM process has defined task content (what tasks?), and estimates of task frequency (when done?) have been made, such that the ideal PM program in Figure 2.1 has been specified. PM Task Packaging must now be completed to take the optimized PM program to the floor.

This seemingly straightforward action has, however, proven to be a difficult step for most organizations to complete in a timely fashion. This situation was, quite honestly, a complete surprise to the authors since we perceived the most difficult hurdle would be to complete the systems analysis process. At first, the reasons behind such difficulties were a mystery because implementing PM tasks (irrespective of their origin) seemed to be an activity that was an inherent part of a plant's organization and infrastructure.

As it turned out, this was not necessarily so, and some of the more important impediments that emerged are discussed below.

8.1.1 Equipment-To-Function Hurdle

Virtually all maintenance personnel, management and technician alike, are consciously (or unconsciously) wired to the concept of "preserve equipment." That is, their very first thought immediately focuses on hardware (equipment) whenever the issue or discussion turns to the subject of "Improvement." This is a very natural reaction since the dominant maintenance philosophy has been rooted in the keep-it-running mindset for at least the past 50+ years.

Along comes RCM. And it says not to think like this anymore. Rather, RCM thinks "Function," and says that this is what must be preserved. Wow—what are they talking about?

This is one of the major hurdles that needs to be crossed. Why should an expert in equipment maintenance be expected to have the reason for his/her job shift from "preserve equipment" to "preserve function"? Here, we are clearly into the buy-in and training issues which are discussed in Secs. 9.1 and 9.4 in some detail.

8.1.2 Organizational Hurdle

There are four areas where organizational factors influence the entire approach to RCM, and these factors are likewise important to how any RCM program is finally carried to the floor. These factors are:

- the Structure Factor
- the Decision Factor
- the Financial Factor
- the Buy-In Factor

Please refer to Sec. 9.1 where each of these factors is discussed in detail.

8.1.3 Run-To-Failure Hurdle

Picture this scenario. You are a skilled technician in every aspect of maintenance for the widget. In fact, you have been recognized by your peers as "the expert" on the widget, and management has rewarded you for this via various promotions and rewards. Again, along comes RCM. And because the widget, in certain applications, has become a non-priority item because of function considerations, some of the widgets have been designated RTF.

This is not a fictional scenario. The authors have encountered situations where air-operated valves have been required to meter flow to critical power plant production functions, and the same valves have controlled flow to the plant auxiliary functions (such as cafeteria, rest rooms). So RCM says RTF on the latter function. Now what do you, the expert, do?

The expert in question, we find, initially has a very difficult time with this problem. Such experts are dedicated to their trade, and proud of their skills and recognition for this. In one actual situation, we found that one of these expert technicians was coming to the plant every fourth Sunday (i.e., on his own time) to continue the PM on "his" widget. It took about six months for this to come to light. The point here is that the RTF hurdle or shock can be very real, and it is paramount that the plant staff be given every opportunity to learn about and accept the first principle of RCM—"Preserve function." If we can handle the equipment-to-function hurdle above, it will automatically preclude this RTF hurdle.

8.1.4 CD and FF Hurdle

Many (if not most) organizations are still locked into the old traditional mode of time-directed (TD) tasks—even those TD tasks that are highly intrusive in nature. Along comes the RCM results, and they are loaded with CD and FF tasks. Worse yet, these CD and FF tasks often bring to light in a very visible and forceful fashion the need to involve the operators as the most efficient way to incorporate their implementation in the plant PM program. "Horrors! Operators doing PM! Not on my watch." as one Operations Superintendent explained to the Plant Manager. Also, if you happen to be one of the fortunate individuals who have been assigned to the application of predictive maintenance technology by a forward-looking boss, you are painfully aware of the resistance that is encountered with your peers.

Change will not occur easily, especially when it requires people to learn new skills and/or perform new and difficult tasks in their job responsibilities. Resistance to change happens to be the nature of the human beast, even when it is patently clear that the change is in the best of interests for him and his associates. We believe that this hurdle is slowly disappearing, but not as quickly as it should. Communication and training again emerge as the most viable solution path for management to pursue. And remember, introduction of non-intrusive PM tasks (Sec. 1.5) is what we believe to be one of the key elements in a World-Class maintenance program.

8.1.5 Sacred Cows Hurdle

The problem here is that certain time-honored PM tasks frequently are treated as untouchable—i.e., sacred cows. The reasons behind this are usually quite rational on the surface. For example, experience shows that "it works" (even though it might be very costly), so why change? Or, it may be that the Regulators have said "do it," and we certainly can't change this—or can we?

Our recommendation is that any and every RCM program should unequivocally forbid sacred cows. The reasoning for this is quite simple. Let the chips fall where they may, and if they should happen to fall on one or more sacred cows, so be it.

But now you at least know that the RCM process has led you to conclude that there appears to be a better way to do it. You still have the option to disregard the RCM solutions, and keep what you already have. But now you also have another choice to consider, and with considerable backup as to how you reached it. You can make a more informed decision, for example, about challenging the Regulation edict if you really believe that the Regulators were wrong in their selections. As a case in point, many organizations are now doing this on a selected basis, and with successful outcomes.

8.1.6 Labor Reduction Hurdle

Discussions about labor adjustment deal not only with the issue of skill and training needs, but also with the question of workforce reductions at the plant. The reasoning goes along the line that if the RCM program is, in fact, capable of reducing maintenance costs on the systems where RCM is applied, then this must equate to a loss of jobs. To date, the evidence that we have observed overwhelmingly indicates that RCM-based reductions do not occur. There are several reasons for this, but two factors seem to dominate. First, a major fraction of any PM savings that may be realized develops from either the extension or the elimination of complex overhaul or repair tasks—tasks that are usually performed by vendors or at outside maintenance shops where plant staffing is not directly affected. Second, plant staffing is usually on the "mean and lean" side to start with; thus, the RCM program is aimed at obtaining the best possible productivity and plant availability for the costs incurred—not the reduction of the staff required to achieve this objective. And third, we have yet to witness any reduction in force (RIF) scenario that was related to any specific productivity improvement effort. Rather, RIFs are essentially directions from higher level management that are driven by the need to reduce expenses by X%, and manpower reductions are one of the easiest ways to do this. The main concern here, if any, should be the possibility that new skills may be needed and the current staff may need to learn these skills or be bypassed as progress marches on. The continuing bent in corporate America is "right sizing" (a euphemism for staff reduction), but these actions are aimed at trimming fat and duplication of effort from *COST* centers, not muscle from *PROFIT* centers. The RCM program deals exclusively with the muscle in a center contributing to the company's profitability.

8.1.7 PM Task Procedures Hurdle

In every RCM program, there are bound to be new and modified PM tasks that will require the generation of new or modified procedures before the task can be performed on the floor. The degree of formality and detail required of these procedures will vary considerably from plant to plant, with safety systems/equipment in a nuclear power plant probably representing the extreme requirement for such formality and detail. The problem encountered here is, quite simply, to determine who is responsible for preparing these procedures. Or, if responsibility can be clearly defined (which is rarely the case), is there time available to write the procedures (in between all of the "firefighting" that occupies such a large segment of the available work hours)? Thus, an RCM program plan and schedule must address, from the outset, just how this will be accomplished. You don't wait until the last minute, and just dump it on some unsuspecting group or individual. That approach usually does not work! Ideally, the plant organization has an identified group whose job is procedure upkeep, and the procedure issue will not become one of those unfortunate stumbling blocks. If this is not the case, then the systems analysts (with appropriate help—externally supplied, if necessary) are probably the next best course of action. The worst thing you could do is to dump the job on the maintenance supervisors and/or leads, who are probably not inclined to do such work or, if they are, will not have the time to do it. (For a discussion on how to develop PM task procedures, see Sec. 8.4).

8.1.8 Labor and Material Adjustment Hurdle

An RCM program typically can be expected to introduce new (and often more technically sophisticated) CD and FF tasks, and also to modify existing tasks, which frequently is the extension of intervals. Both of these changes will introduce labor and material adjustments to the current PM program. Labor adjustments will come primarily in the form of new or revised skill requirements for the craft personnel and technicians to carry out the CD and FF tasks (such as operator involvement). Material adjustments may come in the form of new equipment and tool requirements, as well as a decreased spare parts inventory as the effect of fewer component failures accrues over time. Here again, the effectiveness aspect of the decision process should help to guide decisions on the commitment of capital expenditures where such are warranted (e.g., to purchase a new thermal imaging camera), and the positive effect of reduced inventories should be tracked and measured to reflect the resulting cost savings. The cost-benefit decisions should be quantitative in nature, and not relegated to the gut-feel and "I think" inputs which all too frequently are the modus operandi within today's business climate. Thus, the need to introduce new tools, equipment, and skills should never become an impediment when the return on such investments is clearly beneficial to the bottom line.

8.2 GEARING FOR SUCCESS

Gearing for success in a RCM program is not unlike "success planning" for any other worthwhile endeavor. You have to make the effort to see the whole picture. First and foremost you must plan before you can ever consider how to execute your plan, verify if the results you are getting are those you expected, or consider how to revise your plan to get back on course—always looking to the future and what may lie ahead, and then do it all over again. Those of you familiar with the emphasis on Quality that began in the mid to late 1980s will have in all probability recognized a

Figure 8.1 The Deming quality wheel (PDCA).

close similarity between the RCM success planning process we have just described and the quality process made famous, first in Japan and then throughout the world, as the Deming wheel—Plan, Do, Check, and Act, or PDCA, see Figure 8.1.

Starting in the upper left quadrant of our wheel, shown in Figure 8.1, our successful RCM project begins to take shape.

8.2.1 Plan

First-class RCM project planning begins with asking and then finding the answers to the following four questions:

1. What is your end goal?
2. What resources (manpower, materials, tools, commitments, buy-in, money, meeting rooms, and computers) will you need?
3. How will you secure and keep these resources?
4. Foresee, as best you can, what hazards and obstacles could lie along the way? (e.g. competing fiefdoms, availability of resources, changing commitments, lack of ownership)

Let's address these questions collectively and suggest some answers.

Simply put, you have to determine where it is you wish to be when the journey is over. Is this just a trial balloon, to see if RCM is for you, or have you decided on using RCM as one of the primary tools in developing your new and cost-effective PM program? Either question requires that you first establish understandable and achievable goals. That is not to say that some or all of these goals may not change over time, but you must develop a viable vision of the future—you need to have a plan! A good plan will consider all aspects of the project from its beginning, to its completion, and beyond. The beginning may be a pilot RCM project, and if so, here are a few questions you should ask yourself. This list is, by far, not complete but it should get you thinking, and that is what good planning is all about.

Planning:

- From whom and at what level of management do I need to obtain approvals? Remember that RCM must be supported from the very top of any organization if it is to be successful and have a long and productive life expectancy.
- Where will the funding come from? You need to plan for:
  —Computers and software
  —Consultants
  —Salaries for the company people involved in the project (craft, facilitator, instruction/procedure writers, planners, etc.)
  —Purchase of new equipment to support the methods recommended by the RCM analysis, e.g. vibration or thermography equipment
  —Binders, supplies, and miscellaneous materials
- Who is the RCM champion? *This question is usually connected with obtaining management support. A recognizable champion with a close and positive relationship with the troops must emerge during the pilot phase of the planning process. It is this person who will carry the RCM torch and light the eternal flame. No RCM champion—no lasting RCM program.*
- Do you need the approval and support of other groups, organizations, or individuals outside your direct control? *This is critical, especially if you require use of resources that they control.*
  —Operations
  —Other support groups, such as PdM groups
  —Planning and work control—CMMS
  —Administrative types, e.g. budget, purchasing agents
  —Clerical staff
- How will the recommendations be implemented? *This has traditionally been the stumbling block, the difference between a successful RCM program and one that is not.*
  —Do you have an implementation plan?
  —Can your team directly implement all the recommendations?
  —Who else has a role in implementing the recommendations, and how will you obtain and then sustain their support?
  —What is the impact on your CMMS?
  —How will IOIs (Items Of Interest) be assigned, progress tracked and reported, and the ROI (Return-On-Investment) calculated?
  —What is the impact on other company systems, e.g., stores, accounting?
- What are the potential organizational impacts when this program is implemented? Who will do what tasks—e.g. will operations be expected to do maintenance and will they do it? Are there fears of staff reductions?
  —Plant staffing levels?
  —How is the plant organized?
  —How will the work be assigned?

- How will you measure success, and, very importantly, what are the metrics to be employed? Below are some suggestions (more on this subject can be found in Chapter 10—The Living RCM Program):
- Short term
  —Progress reporting on specific RCM analyses
  —RCM systems completed
  —Changes to PM program implemented
  —Preliminary ROI (Return-On-Investment) from IOIs (Items Of Interest), especially any high impact or dollar findings.
- Long term
  —Change in the total maintenance cost—anything down is good
  —Change in the forced outage rate on RCM versus non-RCM systems and even components
  —Any CM events which RCM missed, and what is being done to ensure it will not happen again
  —Long term ROI, both measured and anticipated from implementing IOIs
- How will you keep management informed as to the progress being made, and how will it be accomplished? *The more visible the program is, the more likely it will continue to be supported—get on their weekly agenda. Also, don't be afraid to report improvement opportunities (e.g., problems)—just have a suggested solution handy.*

The last two items in this short list, metrics and reporting, are the primary keys to the success of any program. We have found that some form of progress reporting is generally quite effective in the near term. It can focus on the number of sub-systems or systems completed. Reporting must include any big Items of Interest (see Sec. 5.10) that the team uncovered—remember that these are pearls of great value. As your RCM program gets more entrenched in the daily life of the plant, the more important it is that you report on metrics that have a measurable value, especially ones that indicate the dollar impact on the bottom line that the program is making. It is a little akin to preaching to the choir, but soft metrics, like avoided cost and other metrics that have a lot of "trust me" in them, do not float the ship. Seek an alliance with your financial group to establish metrics, such as the overall impact on the *Total Cost of Maintenance*—it is the direction of the bottom line, i.e., profitability, that is important. By involving finance early on, you have gained a tremendous leverage, and your reporting metrics will not be questioned.

As you can see, each of these questions begs yet more questions. You may not be able to plan (i.e., anticipate) for it all, but you need to try. The more complete your planning before the project is started, the better your chance of approval, including buy-in and acceptance of the new maintenance paradigm from both management and those who must implement it. Before we move on to the next spoke on the wheel—"Do"—a word to the wise. Any impact, be it direct or indirect, on the resources or fiefdoms of others outside of your direct control must be foreseen, and the necessary commitments obtained; or the lords of these fiefdoms will lie in wait like a sleeping dragon, rise up, and devour you and your project. A word to the wise!

8.2.2 Do

This part is rather simple—just do it. Do what you planned, and keep doing it. Leave no stone unturned, no deed undone, no report not made, no buy-in or acceptance not gathered, and miss no opportunity to show what's been accomplished. Now, carry your plan to the floor, then step back and make sure that it is hitting on all cylinders—being ever so mindful of those dragons who lie in wait.

A word here on implementation, which is part of doing. For those embarking upon their first RCM program, the caveat is quite clear—you must decide on how you will achieve Task Packaging from the outset, plan accordingly, and then proceed. Otherwise, all of the effort and expense incurred in the systems analysis process and the valuable results obtained might become just another dust collector on the shelf.

One solution approach is quite simple—implement as you go, system by system. As cliché as the saying might seem, it is true—Success breeds success! Once positive changes, especially PM improvements, have gained momentum, they are difficult to stop. So remove the opportunity for failure, get the ball rolling early and keep it rolling; do not wait for the big push at the end. This simple suggestion may indeed be the difference between success and failure.

8.2.3 Check

After planning, the second most important consideration in gearing for success is checking. Checking takes place constantly and throughout the entire RCM process, and is yet another series of questions. The answers will indicate where you are on your time-line and how successful you have been. Below, we'll provide a few of the bigger questions that you must ask. And there might be others, it all depends upon the honest answers you get to the following questions:

- What do the metrics tell you?
  —Are you on schedule?
  —Are you being successful?
  —Could it be better?
- What does the plant grapevine say about the program?
- Are you implementing, and how well is it being accomplished and received?
- Are you ready for the next series of success actions that were outlined back in the planning stage?

Finally, have you been keeping both management and the troops appraised of progress and achievements?

8.2.4 Act

Now that you have gauged where you are, and whether or not the results have been satisfactory, the fourth segment on the wheel, Act, will be either easy or will present some opportunities for improvement. Here, too, there is a series of questions that you must answer:

- What are the short and long term impediments to success?
- What future success actions need to take place and when?

In other words, if the answers to our *Check* questions have been positive—what needs to be done to maintain the continued success of the program? And, if negative—what do you have to do to get back on track and be successful?

In performing the measurement actions of *Check*, the successful programs review their *Plan*, revise it, and then implement the new plan—they *Act*. As when a wheel turns, you cannot see the individual spokes but only see a blur that appears to be solid. These four simple action verbs—Plan, Do, Check, and Act—are in synergy with each other only when they are viewed as a continuous function. They are dependent upon each other, and so is gearing for success in your RCM program.

8.3 INTERFACING WITH THE CMMS

If the plant or facility already has a Computerized Maintenance Management System (CMMS), then the entire question of CMMS requirements in the RCM program should be a relatively minor or nonexistent issue. But if the converse is the case, which includes those facilities that have chosen to replace/update their existing CMMS package, people could raise this as a stumbling block to beginning a PM improvement program, even though it may or may not be totally true. There is no question that the CMMS facilitates the efficient conduct of a PM program—any PM program. But with an RCM program, the efficient monitoring of CD task parameters (including an ability for automatic alerts), the compilation of PM and CM cost data for benefit analysis, and the tracking of Age Exploration programs, component histories for statistical analysis, and other related program measurements could become cumbersome or nearly impossible if the CMMS is totally absent or in need of modernizing. Thus, a decision to proceed with an RCM program could also involve the decision to update or acquire a CMMS, if such a decision has not already been made. The decision to purchase or replace a CMMS, while fraught with its own difficulties—and not directly the subject of this book—should never become a reason for not embarking upon an RCM program. In this section, we will constrain discussions to presenting what RCM may require of the CMMS, along with providing some thoughts on how the CMMS should support ongoing and future RCM activities—especially those of the RCM Living Program.

There are two distinct reasons for interfacing a CMMS with an RCM program. The first, and most fundamental, is to download the component list, their associated tags/IDs, and any useful failure history into the RCM analysis software, or to just simply provide that same list to the RCM analysts. The second and far more important reason for this interface is to make seamless and direct use of the invaluable data from the RCM analyses. That is, to get the failure mode, cause, and task data from the analyses into the CMMS, then make use of that data in developing and scheduling work orders, and analyzing CM events to continually improve the PM program. Significant investment is made on improving a PM program and on the CMMS; should they not communicate?

It should be pointed out at this juncture that, as easy as the foregoing appears, there are some major interface hurdles to overcome. For example, as was pointed out in Chapter 5, how the plant is divided into systems for RCM analysis does not always fit seamlessly in the system boundary and description breakdowns used in the CMMS. In general, this disconnect and many others that would make the power of the CMMS available to all users can be traced to the difference in time between installation of the CMMS and startup of the RCM or PM Improvement Program. However, it has been the authors' experience that these disconnects are more likely due to a lack of coordination between the plant "Systems" group responsible for computer applications, the Maintenance department who uses the majority of the CMMS work scheduling functions, and the technical/engineering staff who are responsible for the reliability and availability of the plant's systems and equipment. If you are fortunate enough to already be doing RCM or thinking about starting RCM prior to a CMMS installation, either new or a replacement, you have a golden opportunity to influence the structure, use, and capabilities of the CMMS to support PM improvement activities—Make the most of this opportunity!

The requirements list below suggests a minimal set of CMMS capabilities or functions to support either the simplest or most advanced PM program. The list may not touch on every detail that your particular CMMS or situation may require, but it should go a very long way to getting you started. It does contain all the basics that the authors feel are important. Some of the requirements, especially those dealing with automatic alerts, may appear quite advanced for some situations. They are presented here as opportunities that advancements in CMMS software will include—so you should be ready to use them.

8.3.1 Requirements for CMMS Integrated Support of RCM Activities

1. Ability to identify all assets (components) that use RCM-based PM tasks as "RCM" assets.

2. Ability to identify all PM tasks and/or work orders in the CMMS as "RCM" if the task was recommended by an RCM analysis.
3. Ability to download into an appropriate CMMS file the following data from the RCM analyses:
    * Component (Tag ID # & descriptions)
    * Failure mode(s) and related failure cause(s)
4. Ability to collect/file the same data as described in #3 above from corrective maintenance reports/work orders.
5. Ability to compare data entered into the CMMS from items #3 and #4, and to generate an "ALERT" message (automatically) if either of the following occurs:
    * File data in #3 does not recognize the failure mode and/or failure cause that was entered into the file data in #4 (i.e., the RCM analysis missed a failure mode).
    * The two files recognize each other, but the frequency with which the failure mode and failure cause appear in file data in #4 exceeds a predetermined value (i.e., an RCM decision to RTF misjudged the failure rate and should be revisited).
6. Ability to compare whether or not an Asset that has been involved in a corrective maintenance event:
    * Has an existing PM work order(s)
    * What those PM work orders are
    * And, if the Asset is an "RCM" Asset, provide an ALERT message if a PM work order cannot be identified to exist (i.e., this is an early warning that a failure has occurred in an RTF component and requires re-evaluation).
7. Ability to track data that is being used for Age Exploration and, in particular, to issue an ALERT (automatically) if a predetermined set point or condition has been recorded. *(Such data could come from the craft feedback on a PM work order, or a corrective maintenance action.)*
8. Ability for the CMMS to construct the data needed to establish trends from condition-directed PM tasks (including PdM) and to issue ALERTS (automatically) when predetermined values of the trended parameter have been reached. *(The data entered into the CMMS may be the as-taken measurement, or it may be the inputs from offline data analyses that have been conducted with the as-taken data.)*
9. Ability to provide for any given calendar interval of interest:
    * Cost information for labor and material on PM and CM work orders
    * Summary reports by system or subsystem for Total Cost of Maintenance (PM + CM)
10. Ability to provide for any given calendar interval of interest:
    * Total downtime or unavailability hours for the components, systems, or subsystems of concern
    * Automatically provide quarterly reports by component, system, or subsystem for revenue lost for this downtime/unavailability

CMMS INPUT FORM

Task Title: Task ID:

Requested By: RCM Recommendation: Y   N

Craft:

Equipment(s):   Equipment Tag/ID #s:

Work Task Description:

Task Frequency:

Crew Size / Manhour Total:

Special Instructions / Tools:

_____

Additional Information Required To Implement Work Task:

Entered By: Reviewed By: Approved By:

Date: Date: Date:

Figure 8.2 CMMS input form.

The CMMS input form, Figure 8.2, is provided as one method that the RCM team or PM coordinator could use to transfer the RCM task analysis data to the CMMS. The CMMS Input Form is a simplified version of Figure 8.4, found in the next section of this chapter.

The essence of the foregoing discussion is to provide you with a starting point in relating your PM Program Improvement activities to your daily business routine. Individually, CMMS and RCM represent a sizable investment for any organization—so, it makes sense to maximize the benefits that both provide when they can operate in a mutually supportive role.

8.4 DEVELOPING EFFECTIVE AND USEFUL TASK PROCEDURES

So far, in this chapter, we have outlined several key features for effectively carrying RCM to the shop floor. The final implementation action is to write task instructions and procedures that communicate your knowledge to those who can make use of it, the maintenance and operation crews, and technicians. In the next pages, we will discuss some guidelines for successfully completing the assembly of effective PM task instructions for each component.

8.4.1 RCM Rollup

The first step in implementing RCM decisions is to summarize or rollup the RCM task decisions into a simple usable form that can be applied to the development of effective maintenance and operational procedures.

Using the task decisions made in either the Classical RCM process (Chapter 5), the Abbreviated Classical RCM™ process, or the Experience-Centered Maintenance™ (ECM) process (both found in Chapter 7), rollup or summarize all of the failure modes and their failure causes for <u>each numbered component</u> (i.e., component #) where a task other than RTF was assigned. The left-hand side of Figure 8.3 provides a suggested list of what rollup information should be listed. Note that items 1 through 6 can readily be obtained from RCM Step 5.2 (FMEA), and items 7 and 8 can be taken from RCM Steps 7.1 and 7.2 (Task Selection and Sanity Check, respectively). The right-hand side lists additional facility/plant information that links the two sets of information together. Items 14 and 15, while optional,

RCM Analysis Information	Facility/Plant Information
1. Component #	9. Facility/Plant System Name & ID
2. Component Description	10. Plant procedure # or Job Plan ID
3. Failure Mode #	11. Plant Asset/Component Name
4. Failure Mode Description	12. Plant Asset Tag/ID #
5. Failure Cause #	13. Responsible Department or Person
6. Failure Cause Description(s)	14. *Optional* – Completion date
7. Task Description(s)	15. *Optional* – Procedure Development Status
8. Task Interval(s)	

Figure 8.3 RCM rollup for component # xx.

would become necessary if effective tracking of an implementation schedule were to be adopted. The information in the RCM rollup should then be further sorted by both the component # and task interval to separate each grouping into useful packages for developing the actual procedures.

One method of developing the RCM rollup would be to use a form similar to the task packaging form, Figure 8.4, that expands upon the information listed in Figure 8.3. The central idea is to collect and then correlate all the information needed to develop, write, and maintain a given instruction in one convenient place.

The form in Figure 8.4 has been found useful by several organizations in developing their PM instructions from their RCM recommendations, and is presented here only as an example to demonstrate some of the points made in this section. This form has been employed most effectively as a transition document to expand upon the RCM or ECM task recommendations for input to the CMMS work orders and/or PM procedures that may be necessary as backup information to a work order. If the RCM WorkSaver® software tool (see Chapter 11) was used, the RCM analysis data from the FMEA, along with their corresponding task recommendations and intervals, can be exported to a spreadsheet for easier manipulation.

8.4.2 Procedural Development

PM task instructions are generally built around a single component and a unique task interval. (Your automobile owner maintenance booklets are usually displayed in this fashion.) The maintenance of component groups that comprise a logical path, train, or instrument loop may also be combined into a single task instruction. In the case where components have been grouped, care must be taken that the work being performed can only be completed effectively when all of the components receive maintenance at the same time. If the task intervals differ, the simple combining of components should not be done. The deciding factor is whether or not operational parameters of one component are hard linked to another, so that servicing one component requires the servicing of other components (this is the theory behind the calibration of instrument loops).

Once the component and interval upon which the PM is to be scheduled have been determined, the individual tasks are then aligned in a normal sequence that ensures all parameters are checked or performed in the proper sequence. As previously mentioned, several components may be grouped in a single PM, but consideration needs to be given to who does the work and the type of task (calibration, overhaul) to be performed. If the work is multi-disciplined it may be better to write a separate instruction for each craft group, especially since individual groups usually control the issuing of their own PMs.

However, coordination between the crafts should be part of each instruction; e.g., if a mechanical task to remove a valve requires an instrument technician to first

Task Packaging Form

Job Plan ID: _____   Date: _____

New _____ Existing _____

Facility/System Name & ID:_____

Component Name:_____ Tag/ID #:_____

Task Interval /Frequency:_____ RCM Completed: ___Yes ___No

RCM Analysis Information
Title:_____ Date:_____
Facility:_____ RCM System:_____
Subsystem:_____ RCM Component #:_____
Responsible Department & Person:_____

Craft: _____ *Man Loading:* _____   *Estimated Manhours:* _____

Engineering Analysis Required:_____ For What:_____

Major Task Elements: (List RCM Failure Mode and Failure Cause #'s)

1.
2.
3.
4.
5.
6.
7.
8.
9.
10.

Special Instructions:

Special Tools:

Required Parts:

Figure 8.4 Task packaging form.

remove the valve's controller, then this is written as a hold-point in the mechanical instructions and a separate PM to remove the controller is issued by the instrument group. As already indicated, the use of a form similar to Figure 8.4 can assist the procedure writer to develop clear and concise instructions to implement the information contained in the RCM analysis.

1. Instruction Title
2. Task Instruction Number
3. Task Interval
4. Priority
5. Manpower Required
6. Estimated Manhours
7. Actual Manhours
8. Component Name(s)
9. Component Tag/ID Number
10. Coordination (Other groups – contact with telephone number)
11. Special Safety Concerns and Instructions
12. Equipment Tag-Outs or Clearances (Which ones, when obtained, and who granted)
13. Tooling (standard and special)
14. Other Information
15. Task Objectives
16. Detailed Task Steps

    In the step that is specific to a failure mode analyzed in the RCM analysis, include the failure mode number and description and the failure casue or causes in brackets, e.g. [FM 12,.11, Failed insulation, and its FC, Heat and Age] Include special notes, warnings, etc. that enhance knowledge or performance.
17. Post Maintenance Testing Requirements
18. List of As-Found and As-Left Conditions and Measurements
19. PM Effectiveness Questions:
    - "Did the as-found condition warrant the work performed?" Yes/No
    - "Could the component have gone longer before doing this task?" Required immediate action/Required minimal to no action
    - "Could the effectiveness of this PM be improved?" Yes/No and How?
20. Notes and Observations

Figure 8.5 Maintenance task instruction content list.

The authors, as a result of their years of reviewing maintenance programs, have compiled a list of items that were found in the maintenance task instructions of effective maintenance programs (see Figure 8.5). It should be noted that not every instruction requires this level of detailed information. However, those responsible for developing maintenance task instructions should consider each of the items and determine their applicability for inclusion.

All failure modes with a designated PM task in the RCM analysis must be contained in one or more PMs (including operator rounds or checks). Failure modes that were designated RTF or rare event should not be in any of the PM procedures. All PM procedures should have a place to note or record problems or miscellaneous information. Finally, review all comments made in the RCM analysis, looking for special instructions, notes, or warnings that should be included in the procedure.

The details of the "what to do and how to do it" (i.e., maintenance instructions) are not directly a part of any RCM analysis; this information is often taken from vendor manuals and instruction sheets. It is always advisable to consult experts— the craft persons who know what should be done and in what order it is best to do so. In some organizations the craft technicians themselves may be the major developers of PM task instructions.

A word about items 18 and 19 in Figure 8.5. Item 18 suggests that a maintenance task instruction should contain feedback on the *As Found—As Left* conditions observed and measurements taken during the performance of a PM. Experience has shown that each component has a few critical failure mechanisms (i.e., the drivers of maintenance or why maintenance has to be performed in the first place), and if we can determine the extent of their progression to failure, we can gain invaluable insights into the effectiveness of our PM task intervals (i.e., Age Exploration). The RCM analysis helps us to determine these critical failure mechanisms, and by collecting even anecdotal information about them we can improve the overall effectiveness of our maintenance.

Likewise, the three questions of Item 19 tell us if our procedure is indeed properly focused on what does go wrong and even if we are doing the right things at the right time. The answers given to these three questions, when tracked over time, tell us if there is room for improvement in our maintenance—are we at the right spot both in what we do and in the interval at which we do it, or have we gone too far? Together, items 18 and 19 collect vital signs of successful PM programs so they can be monitored by our Living RCM Program to ensure that the benefits from the RCM analyses are realized and maintained. (See Chapter 10 for a more detailed discussion on the Living RCM Program.)

There is no one single accepted procedural format, as it relates to the degree of content, other than the common agreement that it should contain enough information to get the job done safely, in as short a time as possible, without adversely affecting the reliability of the component being worked on or those around it, or the availability of the function it is there to perform. How much is too much is hotly debated, especially from those who envision that their job is to create a risk-free world. On the other side of the same coin, the position of most knowledgeable craftspeople is to have concise and meaningful instructions that avoid lengthy treatises or maintenance philosophy.

Indeed, the truth lies somewhere between the two extremes. The trick is to find an acceptable balance between imparting knowledge and not taking the wrench out of the hands of the maintenance professional. Books have been written on this very subject. However, that is not the intention here, so suffice it to say that to develop an effective PM task instruction, you must understand, as a minimum, the following:

- the skill level of your people (which will change over time, especially as the workforce ages and retires),
- the risk of doing any form of intrusive maintenance (data has shown that about 50 percent of all such intrusive actions result in service re-visit),
- the technical requirements of properly maintaining any given component, and
- the requirements of your managers, regulators, and insurers.

In the final analysis, what makes a PM task instruction effective is the ability to impart just enough information—information that you would not reasonably expect the craftsperson to automatically know (e.g., left-handed threads on a nut, foot-pounds of torque required to secure a special bolt, a specific order of removal, warnings where mistakes have been made or safety impacted). On the other hand, do not repeatedly tell them standard practices which they are trained and certified to know. If we take away the responsibility of performing a task right and in a safe manner, apathy will inevitably creep into our workplaces and we may get what we were trying so hard to avoid.

9

RCM LESSONS LEARNED

It is very likely that most of the people involved in your company maintenance program, especially those charged with its improvement, have heard the term "RCM." However, the number of people who actually have some reasonable understanding of RCM is, in our experience, quite limited. Given that such an uninformed state exists, it is extremely difficult to gain wide acceptance of an RCM program and the value that can be realized from it. The necessity for such familiarization is important at both the management level and the system engineer and craft personnel level. Familiarization at the craft level is especially important, a point that is not always fully recognized or appreciated.

Throughout this book, we have endeavored to demonstrate how RCM can and should be an integral part of any maintenance organization, especially those wishing to be known as *World Class*. In Chapter 8, we have just dealt with how to successfully carry an RCM program to the shop floor and make it the backbone of any PM improvement philosophy. The authors, having been involved with RCM from its first introduction to the U.S. commercial and industrial world in the early 1980s, have experienced virtually every characteristic that contributes to successful and likewise not-so-successful RCM programs. In this chapter, we will attempt to give you the benefits of our 35-plus combined years of RCM experience and the lessons we and others have had to learn along the way. It is our hope that, by knowing where the more significant potholes lurk, you may avoid them.

We will begin our discussion on RCM lessons learned by more fully developing the organizational factors that have an influence on beginning a *World Class* journey in maintenance. From there, we'll touch on what we feel is the proper composition of the RCM team, present some thoughts on how to effectively schedule your RCM activities, touch on the importance of training everyone and not just the RCM team, how to know what systems are the best candidates for RCM and which are not, and how to make the most use of those IOIs. We will then end our presentation of Lessons Learned with winning strategies on gaining acceptance from your peers, program management considerations, and finishing with what we feel are the key factors in the successful and not-so successful RCM programs.

9.1 ORGANIZATIONAL FACTORS

9.1.1 The Structure Factor

There is an old adage that it is the dispositions, personalities, and motivations of the people, not the structure of the organization in which they work, which ultimately determine project or product success or failure. Experience bears this out, in the authors' view. But, by the same token, this experience also says that the particular version of organizational structure that is employed can be a significant factor in making success easy or difficult to achieve. For example, organizational structures usually determine lines of communication, which can be short and simple or lengthy and complex; they also establish boundaries on areas of responsibility which can be either very broad, highly partitioned and restrictive, or even deliberately overlapping and competitive, to encourage the "best ideas" to emerge victorious. (We have worked in the latter organizational philosophy on occasion, and have frankly found it to be quite counterproductive to ultimate product success, and sometimes even destructive of highly competent people who were inadvertently caught in its web.) In the maintenance world, in particular, there is a continuing debate over organizational structure at both the corporate and the plant level that never seems to reach a satisfactory resolution. We refer here to a structure where maintenance and production are separate and equal organizations versus a structure where maintenance reports to production. We have been involved with clients in both camps, and have even worked with clients who have switched from one to the other in midstream of a successful RCM program.

Here is our view on this issue. We see two important factors that should influence the choice:

> a. Externally, meeting customer requirements (delivery, quality, cost).
> b. Internally, achieving team play and efficient use of resources.

We believe that this can best be achieved when maintenance and production are peers—i.e., separate organizations. When maintenance reports to production, team play tends to take a back seat to production's authoritative approach. More bluntly, production loses sight of the vital role that maintenance plays in

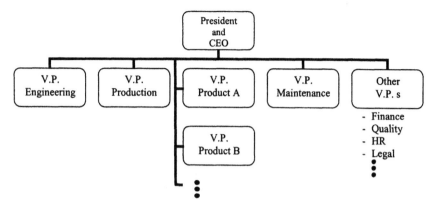

Figure 9.1 Typical corporate organization.

its success; egos and turf battles tend to replace team play. And from an RCM point of view, initiating and implementing an RCM program is easier to accomplish when the two organizations are separate since the decision chain is shorter, straightforward, and more willing to undertake innovations that make the job more efficient. Ultimately, reduction in team play and efficiency will detrimentally impact one or more of the parameters that affect customer satisfaction.

To keep it simple, we will now look at two tiers of a typical company structure that is composed of separate maintenance and production organizations. The corporate level is shown in Figure 9.1 and the plant level in Figure 9.2. How should we deal with them to initiate and implement an RCM program?

9.1.2 The Decision Factor

In the corporate structure (Figure 9.1), our interests usually reside with the Vice President of Maintenance, and also possibly a key technical director who oversees process improvement. And even though the organizations are separate, it is prudent to include the Production Vice President in some of the "selling" discussions because the RCM process ultimately requires the cooperation of Production's equipment operators in optimizing the PM tasks. The important thing is to understand just who the essential decision makers may be; failure to include all of the right people in the selling process could doom your efforts before you even reach first base. One simple example will illustrate this latter point. You manage to do a first-class job convincing the VP–Maintenance that RCM is needed. But nothing ever happens because, unbeknownst to you, the VP had to then sell it to VP–Production who controls approval rights to any production line modification, and your VP flunked the course. You might well have succeeded had you made the pitch to Production, but you never knew about the control that production exercised on new maintenance ideas.

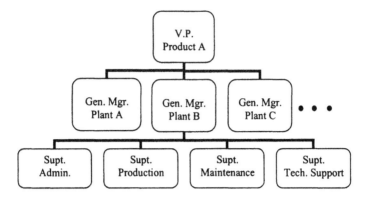

Note: In a typical matrix organization, the plant personnel are
        drawn from the parent Functional organization.

Figure 9.2 Typical plant-level organization.

In the plant-level structure shown in Figure 9.2, our interests deal with the plant general manager as well as the superintendents of operations, maintenance, and technical support. Of course, even to get the plant's attention, we might have first required a positive endorsement from corporate management. Without this endorsement, we may get nothing more than a polite hearing from plant personnel, if that. Notice also, that when dealing with the plant, operations and technical support will play a role that could be just as important as the maintenance role in initiating and achieving a successful RCM program (see Sec.9.2— RCM Teams). If the plant manager happens to be a "power center" on the organization chart, it could be that your sales job begins and ends there. We have seen such cases, but they are infrequent. A good plant manager, however, can be a very strong ally for your cause if he introduces and endorses your idea at the corporate level. In either event, it may sometimes be prudent to start at the plant level even though the real decision authority probably resides at the corporate level.

9.1.3 The Financial Factor

Our corporate focus in the previous discussion on the decision factor is directly related to the financial factor. Funding for new initiatives like an RCM program either comes from a corporate budget, or at the very least may require a corporate-level approval of development expenditures that are resident in a plant budget. You may also find that everyone is apparently positive about the introduction of RCM, but no one is willing to pay for it. We have all too often heard responses to the effect that "we can't afford it." The job then becomes one of convincing the decision makers that they "can't afford not to have it." How is this done?

There are several parameters that influence a credible answer to this question, but the answer in a nutshell is ROI—Return On Investment. What will it cost, and what will be the payback? Let's examine these two questions more closely.

On the cost side, we need to put away one RCM myth that is frequently encountered—namely, a belief that RCM must be applied to every system in a plant in order for its benefits to be realized. And this myth is frequently what is behind the "we can't afford it" response. So let's put this myth to bed right now. You do not want to apply RCM to every system in a plant! Sec. 9.5 will explain in more detail that you usually choose about 20% of the plant systems for an RCM program using the classical process, and maybe another 20% (if any at all) using the abbreviated classical process. Using these percentages, we can suggest the following guidelines for costs (assuming use of "RCM WorkSaver" software—see Chapter 11):

Classical process, using a 3-man team (see Sec. 9.2):

- Pilot (1st) system—about 6 weeks applied time, including training time, or 18 man-weeks of total effort.
- Subsequent systems—using those personnel trained on the pilot system, there is a rapid learning curve to about 4 weeks applied time, or 12 man-weeks of total effort.

Abbreviated classical process—about 75% of the effort required on the full classical process.

The cost of RCM WorkSaver software and use of a consultant for training and pilot project facilitation can add a one-time cost of about $40,000 to $50,000. For a rather simple plant with about 30 systems (e.g., a fossil power generation plant), the RCM program would cost $200,000 to $300,000 over a 1- to 2-year period. If the plant is complex with, say, 100 systems (e.g., a nuclear power generation plant), the cost is in the $900,000 range over 3 to 5 years.* Of course, there are many variables that influence these cost and schedule figures, such as learning curves, personnel experience, salary levels, number of teams employed, and team size. (It is important to note that, when multiple plants or facilities are involved, a new RCM analysis for each system is usually not required; rather, the existing systems analysis can be "replicated" at the other plants at a considerably reduced cost. The replication process is a particularly useful concept where a number of similar plants/facilities are involved.) But at these levels of expenditure, it is safe to say that the approval requirements are at the corporate level where the Vice President of Maintenance, together with key members of his or her staff, must formally concur in order to establish a line item in the budget for the

---

*These cost estimates are given in 2002 dollars, and assume an 80/20 rule using the Classical RCM process.

RCM program. Thus, the initial sell occurs at the corporate vice-presidential level, and this may or may not occur with visible support from the Plant General Manager. If you are fortunate, it is the Plant General Manager who initiated the request for RCM program funding because the selling job is then already halfway done at the outset. We have occasionally experienced this, and it has simplified the entire process immeasurably. In one rather extreme case, the board of directors became the approval authority, but the more general rule is that you must go to the corporate vice presidents to get the ball rolling. If two vice presidents are involved, your selling job may be more than twice as difficult—simply because these people may have different agendas and priorities which tend to compete for available funding. Thus, you may have to gain a very comprehensive understanding of these two agendas, and then find a way to couch your sales pitch to fit both agendas. If a single vice president is the initiating sponsor, funding approval becomes less involved since he or she usually will not hesitate to act unilaterally if you present a convincing case that success is highly probable.

But are the above costs worth it? Again, what will be the payback? Let's start by recalling Figure 1.1 in Chapter 1. Our point there was the need to focus our PM resources (costs) on decreasing the CM incidents (costs) and increasing output (profit) by reducing forced outages (i.e., downtime). This latter factor is by far the big swinger in this financial picture. A brief downtime analysis, based on a very conservative estimate, will place ROI quickly into perspective. We find that all of our clients measure a one-day loss of output in the $100,000 per day range and up—a nuclear plant, for example, must purchase about $800,000 of replacement electricity when it goes unexpectedly off-line for one day. We suggest that a <u>saving of just one day of downtime</u> essentially represents the breakeven point for implementing a comprehensive RCM-based PM program. More realistically, our clients have measured the following benefits (ROI):

- Downtime reductions of 40 percent and up.
- CM cost reductions of 30–50 percent.
- Items of Interest (IOI) paybacks of $100,000 and up.

All of these benefits are annual paybacks. If you agree with these values (or if you wish, only one-half of these values), how can you afford not to implement an RCM program!?

One final thought here. Be sure to check Sec. 12.2 which presents seven case studies that our clients have graciously agreed to contribute to this book. And listen carefully, please, to what they have to say. Also, in Appendix C of this book the reader will find a discussion on ProCost©, a financially based reliability improvement model that can calculate an estimated ROI for commencing and implementing RCM PM program recommendations. A financial model like ProCost© can take the "trust me" out of PM improvement justifications. This model is currently being used by a large Midwestern manufacturing company. Also, we should mention that using financial models like ProCost© adds additional but minimal front-end costs to any project.

9.1.4 The Buy-In Factor

Buy-in is the process whereby an individual or a group, responsible for carrying out some new procedures or actions, has been a party to the development and planning for those actions, and has agreed that this new modus operandi is good for all concerned, and therefore will support its use. When buy-in is successfully accomplished, the people involved have usually made a direct or supporting contribution to the action plan and, implicitly, have accepted the plan as well as assumed some level of ownership in the plan. With RCM, this process occurs almost entirely at the plant level of the organization. Without the essential ingredients of *acceptance and ownership* it is highly improbable that a plant staff will feel motivated and compelled to implement anything—and that especially includes the recommended PM tasks from an RCM program.

Achieving an appropriate level of buy-in to an RCM program is dependent upon several factors that deal with how the plant staff is integrated into the RCM systems analysis process and the expected benefits to be realized. First of all, there must be a clear and visible endorsement for the RCM program from top management—usually at the Maintenance Vice Presidential level. However, do not be lulled into believing that the sales and education job stops there; it does not! If you succeed in obtaining top management endorsement, your job has really just begun because you now must do the same job, only better, with the plant staff. This, in particular, includes the craft technicians who may never be members of an RCM team. This latter process is not a one-stop job, and chances are excellent that the sales and education process will continue over a long period of time—say, two years or more—to capture everyone in the plant organization who is germane to a completely successful RCM program. The approach to buy-in from the plant staff is multifaceted, and requires not only training seminars and one-on-one tutorials but, more importantly, the involvement of experienced and respected craft personnel in the systems analysis process who can explain the methodology and benefits to their peers, and motivate a broad base of acceptance.

For sustained success, an RCM "champion" must emerge who can provide leadership to the buy-in process.

Another important point to consider is the need to include the operations and technical support personnel in the acceptance and ownership issue. The reasons for this are imbedded directly in the RCM process itself, when CD and FF tasks take on equal importance to the more traditional TD (overhaul and intrusive-type) tasks. And who "owns" a share of responsibility for the CD and FF tasks? You've got it— operations and technical support. If you should fail to recognize this facet of the plant organization, successful implementation of the RCM program may never occur.

Our experience is that some organizations do not fully comprehend the significance of the buy-in issue. These organizations typically have serious difficulties when introducing anything new to an operating plant. This problem is compounded when the direction comes from "outside of the fence." While difficult to quantify, we believe that the success achieved with RCM is directly proportional to the degree of buy-in achieved with the plant operations, maintenance, and support technicians.

9.2 RCM Teams

9.2.1 Resource Allocation

Where will the personnel to staff the Systems Analysis Process and Task Packaging efforts come from? This has all too frequently been a monumental issue which has led to delays in initiating an approved RCM program—delays that in a few instances have literally spanned several years. The nature of the difficulty with this issue involves the unfortunate fact that the most logical place to staff and conduct the RCM program, the plant itself, usually does not have sufficient availability of experienced personnel to do the job. The issue of plant on-site staffing is, however, a realistic issue since most plants have had or continue to conduct staff reductions to make them "lean and mean." The plant, without doubt, is the logical first choice for the selection of team members when we consider both the role that buy-in plays in assuring a successful program and the absolute necessity to establish a knowledgeable equipment and operations "database" for the systems analysis process. There are four possible solutions to this issue, each of which has been employed at one time or another to overcome the staffing dilemma.

1. Bite the bullet and assign appropriate on-site plant personnel to the RCM team by giving it top priority over other activities. The problem with this solution is that the top-priority assignment frequently goes by the wayside when there is any sort of hiccup in the plant availability status. Personnel are continually pulled away from their RCM team assignments "for just a few days to handle the crisis" with the net effect that a smooth and continuous RCM program is difficult to maintain. Everyone gets frustrated in the process and, in the worst case, the program may never be completed while, at best, the schedule for the program can be extended far beyond the original target dates for implementation. If you can eliminate, or at least consciously minimize, the RCM team disruptions due to emerging crises, this can be a very effective solution.
2. A variation on this theme is to authorize an increase in plant staffing specifically to assume the normal workload of the RCM team members. With this approach, some (one or two) key personnel from the existing staff might be placed in lead positions to help orient and integrate the new personnel into the plant community. The use of "retirees" has

proven quite effective in performing this "fill-in" assignment since they already are familiar with the plant and its equipment. There have been cases where this approach has worked exceedingly well, and has ultimately produced some of the best implementation results while avoiding any disruption to the daily work routines.

3. A third possibility is to staff and conduct the RCM program through the technical support group at corporate headquarters. This is often considered to be the best solution from a staffing point of view, but it also turns out to be a poor solution in terms of the required buy-in at the plant. This latter point can be mitigated to a large degree if the corporate-staffed RCM team plays a continuing and highly visible role with involvement from and integration with the plant personnel. Success with this approach is totally dependent upon how well this "if" is handled, but it still retains some of the disruptive problem inherent to #1 above.

4. A fourth approach is to bring in an outside contractor to execute the entire RCM program. In this case there is usually some degree of plant management assigned to oversee the contractor, with the net result that a very minor participatory role is played by the plant staff. This condition not only creates a major buy-in problem at the plant, but it may also create some technical deficiencies in the systems analysis process, since the contractor personnel will not have the in-depth systems and equipment knowledge required to thoroughly perform this process. The results of this approach have been mixed. We hear claims of successful programs, but more often we have known of partial to total wipeouts (i.e., the contractor's product was partially or totally unusable). This approach is usually the most expensive one, and has the lowest probability of yielding completely satisfactory results.

A variation of the contractor theme has been employed by several companies with whom the authors have worked. The examples presented in Chapter 12 are illustrative of this theme, where the company has used one of the first three approaches just described, and has employed a single consultant to work with the RCM teams until they become RCM process experts in their own right and can then complete the program using in-house personnel exclusively.

In summary, it is the authors' view that approach #1, augmented during the early program phases by an expert outside consultant, offers the highest probability of RCM program success. We would discourage any consideration of approach 3 or 4—both of which have a high probability that a successful and comprehensive RCM program will never be achieved.

9.2.2 Team Makeup

The team should comprise no more than 4 to 5 people plus a facilitator. Larger numbers definitely fulfill the old adage that "too many cooks spoil the soup."

At a minimum, the team must contain an operator, a mechanical technician (machinist), and an electrical/I&C technician. The well-balanced RCM team is mostly composed of craft personnel representing both operations and maintenance. The maintenance technicians know the equipment inside and out and how the equipment degrades and even fails, while the operators know how the plant systems interact and functionally behave. Most engineer types are only vaguely familiar with these details. Occasionally, a maintenance and/or system engineer is also a team member if this person knows the plant. Historically, however, teams that do not have craft personnel as members have not achieved a successful RCM analysis. We feel so strongly about this latter point that we will not facilitate any team that does not meet this criterion.

9.2.3 Personnel Selection

Not everyone has the temperament and motivation to participate directly in an RCM process. Thus, in selecting team members, it is advisable to choose people who are capable of contributing new ideas, can accept change from the "old ways," and have a desire (thus the motivation) to play a role in improving how business is conducted. Frankly, if people have little creativity in their daily activities and responsibilities, they will not have a positive influence in a "team" environment. This particular team makeup is considered necessary for both the Classical and the Abbreviated Classical RCM™ process as well as for the ECM process (as described in Chapter 7).

RCM team assignments, especially for pilot RCM projects, should not be seen solely as a training exercise. Success is very dependent upon the knowledge of the individuals who comprise the RCM team. However, the assigning of an additional person to "experience and learn" can be an effective training scenario—just do not overdo it.

9.2.4 Facilitator Role

Ideally, a positive team environment greatly helps any new process to be a success. Thus, some degree of a prior team Buy-In to the RCM process will help to create a conducive environment where each member has no reluctance to share his or her experience, knowledge, and opinions with others. In reality, this Buy-In may not always exist at the start of the project, and the Facilitator needs to provide the necessary guidance to help in achieving Buy-In during the early stages of the project. In an extreme case, where a team member consistently takes a negative posture, it might be best to replace him or her. (While this is not the norm, it has occurred.)

A successful RCM project depends on the capture of the team's past operations and maintenance knowledge and experience into the structured format used in the RCM and ECM process (see Chapters 5 and 7). However, the Facilitator must be on guard to assure that this experience does not dominate the process to the extent that few, if any, "new" ideas are introduced. This is especially crucial in Step 7-1 where the team is required to specify candidate Applicable PM tasks for the critical failure modes. Promote innovation and new ideas, even if they go against the traditional way of doing things. In fact, a good Facilitator will go out of his or her way to encourage the introduction of new and innovative methods and techniques. RCM recommendations present strong and defensible *Business Cases* for new or improved cost-effective PM tasks, especially CD tasks.

On occasion, a team member is reluctant to speak out—especially if the team discussion revolves about some controversial issue. It is important that the Facilitator recognizes this, and tries to use his influence to persuade this member to "open up" and contribute his or her expertise more fully.

9.3 SCHEDULING CONSIDERATIONS

Successful RCM teams are always composed of personnel who are known to be among the "best" from the available candidate list. For the RCM team, that is the good news. The bad news is that these very same people are in demand, and represent key resources to management with limited availability for peripheral assignments. This has been a common problem with all RCM projects.

The solution that has worked successfully involves the use of a staggered calendar schedule for the RCM team—generally one week of effort on the RCM project and then a two- or three-week interval where the team personnel return to their normal job duties. This arrangement satisfies most concerns about conflicting priorities, allowing plant management to effectively schedule these "key" personnel. But it is imperative to secure a firm management and team commitment to the one-week intervals when the team personnel will, in fact, be available without interruption for the RCM project. Intervals shorter than one week at a time make it very difficult to complete the analysis work with a reasonable degree of continuity.

When first introduced to management and those responsible for getting the work done, the idea of releasing their best people for one week without interruption appears to be at best somewhat unrealistic and at worst a nightmare. But consider this scenario:

> *Your improvement team has finally found an opportunity to meet for one or two days. By the time you get everyone's attention and redirect them to the purpose of this meeting, you are out of time and little if anything has been accomplished. What a waste of valuable time and talent! Now imagine how much more productive and how much better the results would be if you*

could keep your team assembled for five straight days and concentrate all of your resources on the assigned task.

As we have shown, RCM is a paradigm shift in attitude and philosophy and it takes time to get the ball rolling—especially with people who may not be used to sitting in a meeting all day—so anything that affects their effectiveness will just short-circuit the results. The team, especially the craft members, need to be assured that they will be left alone so they can concentrate on their valuable task—improving the profitability of your maintenance program.

Going hand-in-hand with making effective use of your resources, another reason for the one-week schedule is the cost-effective use of consultants. It makes for better use of their time and expense, especially with today's rising travel costs.

Facility "scheduled outages" tend to play havoc with the scheduling of the RCM team. Consideration of these scheduled events should be taken into account when the RCM team meetings are planned. Far more disruptive are the unplanned and therefore unanticipated outages. These events drain available manpower to get the plant back up, and those individuals most valuable to plant restoration are the same people who were felt to have the "right stuff" and were assigned to the RCM team. From a practical point of view, these interruptions cannot altogether be avoided. It is hoped that the importance of completing the RCM assignments, thus improving the bottom line and decreasing the occurrence of these unanticipated outages, will be recognized. Hopefully, management will become sensitive to this problem, and take action to accommodate at least some of these perturbations in order to keep the RCM project schedule on track.

We have come to learn that the successful completion of the first (pilot) RCM project tends to reduce, if not eliminate, the concerns that first arose over the assignment of key plant technicians to the RCM team.

9.4 TRAINING

Not everyone needs to become an RCM expert in order to realize its benefits and support its introduction. However, it is critical that everyone, from the corporate level VP of Maintenance (or equivalent) down through all levels of plant management, and on to the craft technicians who will be asked to utilize the RCM recommendations, becomes aware of what RCM is and is not. Specifically, focused indoctrination training, about 4 to 8 hours, should be provided to all corporate and plant personnel who may be affected by or have an impact on the program.

It goes without saying that all RCM team members, and especially any new or replacement members, must be trained in the details of doing RCM. It is imperative that all team members have knowledge of the RCM process and how it is employed. This training includes:

- Understanding that the current maintenance situation can be improved upon.
- What RCM is.
- How RCM will help plant management and craft alike to achieve a more cost-effective PM program, and ultimately lead to more personal satisfaction in their job.
- A detailed explanation of the 7-step RCM process, i.e., how to do RCM.
- What the team's roles will be during the analysis and implementation processes.

This extensive training for an RCM team is most effectively accomplished in a 2-step program: (1) classroom-type instruction for a 3- to 4-day period at the outset of each new RCM project, and (2) hands-on involvement in an actual RCM project under the guidance of a skilled RCM facilitator. This 2-step program should be continued whenever a "new" team is formulated to conduct an RCM analysis. Historically, classroom training alone has been tried, and has not worked. Hands-on experience under a skilled facilitator is needed to realistically qualify an individual as an RCM "expert."

If the RCM philosophy is to be institutionalized in the plant, Facilitators must be developed and must have specialized training on reliability and failure concepts, RCM processes, running a team, and efficient use of RCM-oriented software (see Chapter 11). It is very important that an RCM Facilitator be included on as many teams as possible before letting them act autonomously; even then, from time to time the RCM Facilitator should sit-in with all teams. Several of our clients have used consultants who will first train the Facilitators and then counsel/tutor them as they direct a team.

Thus, the issue of training ranges from a broad-based indoctrination program to a focused and intensified program for those personnel directly participating on the RCM teams. Hands-on training via project participation (versus classroom-only training) is clearly the best, if not only, way to conduct a successful RCM program.

9.5 SYSTEM SELECTION

We believe that one of the five ingredients required to bring your maintenance program to World Class status is to "focus resources for the best Return-On-Investment." In order to do just that, it is necessary to apply a credible method that will provide this focus. That method, in our view, is the 80/20 rule previously discussed in Sec. 5.2. We reiterate the 80/20 rule here to again emphasize its importance in any maintenance optimization program.

Hard (and embarrassing) early-on experience has taught us the need to use the 80/20 rule. In two of our early RCM efforts, the clients elected to employ qualitative (i.e., judgmental) decisions on where to focus resources (i.e., RCM projects) to improve their maintenance program. In both situations, the decisions initially led to the selection of systems for RCM evaluation that were "well-behaved" systems with no ROI potential.

So, if you wish to improve your maintenance program—by whatever process— we believe you must use the 80/20 principle as the starting point.

Our use of the 80/20 rule has consistently provided a credible basis for use of the Classical RCM process, and has been a very effective tool for defending the specific system selections that were made. By requiring the use of quantitative data and Pareto diagrams, we have also avoided system selections that gave the appearance of valid 80/20 systems but were totally ill-suited for the RCM process. Two examples in this regard will illustrate this latter point. In the first instance, the system was a high-cost maintenance system during the 18-month period that was selected for evaluation—but the maintenance problem resided almost totally in a single assembly which had been recently replaced with a new design. The maintenance problem vanished, and the discovery of this situation was revealed during a presentation to system engineering management who were responsible for approval of system selections for the RCM program. In the second instance, a high-maintenance system was correctly selected per the Pareto analysis— but a closer review of the system revealed that it was almost entirely digital electronic equipment. In case you haven't already noticed, preventive maintenance on digital electronics is virtually non-existent (do you perform PM on your TV set?). In both of these examples, we would have eventually discovered the problems described above, but some careful review of the selections made in Step 1 of the system analysis process can avoid some costly wasted effort (and perhaps some unwanted embarrassment).

As a reminder, we have found that there are three primary sources of historical system data that are suitable for the 80/20 analysis:

1. Total (PM & CM) maintenance cost.
2. Forced Outage Rates or Downtimes.
3. Number of CM events.

These items should be evaluated for the most recent 12–18 month interval for which data are available. We have also had situations where all three items were available, and each provided essentially the same list of 80/20 systems but in a slightly different order. We have also found that most organizations have records on all three items, but a simple count of CM events is usually the easiest and quickest data set to retrieve and evaluate.

9.6 Using IOIs (Items of Interest)

The introduction of IOIs to the system analysis process was an innovative addition that we first employed very early in our RCM work. When we learned to use RCM teams that were composed of O&M craft technicians, it was quickly recognized that the depth of practical talent gathered about the table, in conjunction with the depth of discussions triggered by the 7-step systems analysis process, exposed large amounts of valuable information above and beyond just maintenance data. So we instituted the IOI list in order to capture these pearls of wisdom.

The IOIs represent an invaluable "free" source of potentially large cash paybacks. For that reason, it is wise to selectively recommend some IOIs for immediate evaluation and action. Our experience is that these early IOI actions frequently produce cost savings that literally pay for the entire RCM program even before the first pilot RCM project is completed and implemented. This early payback feature has always led to a very positive response from management.

9.7 O&M Peer Acceptance

A special sub-category of buy-in deals with the issue of peer acceptance. Every organization structure places each individual in a position where there are peers— i.e., people with virtually the same level of responsibility, salary and, to some degree, influence on how people of "equal rank" might respond to new ideas and changes to the status quo.

Your peers, however, are not always in your part of the organization structure. The maintenance people are, for example, in this situation with respect to their peers in operations. Historically, in fact, maintenance and operation technicians have been at odds for as long as we can remember—with each blaming the other for almost every plant problem that occurs. In a *World Class* scenario, this hostility must cease. Understanding and cooperation between operations and maintenance must be a way of life.

RCM, because of its focus on the necessity to maintain function and its unique approach to the team makeup, will break down many of these barriers by exposing both parties to the everyday trials, tribulations, and responsibilities of each other. Our experience is that RCM has been the major influencing factor in those organizations where the traditional roles of O&M have been successfully melded together. The message here, then, is that O&M personnel need to learn and appreciate the mutual dependence that they share in achieving a *World Class* status.

9.8 Program Management Considerations

9.8.1 Feedback to Management

Management has a large and constantly changing agenda with which to deal. Their nature is to move on and focus on emerging topics and/or the crises of the moment. It's not that they have forsaken you, but that you may have dropped below their radar screen. (In this instance, absence does not make the heart grow fonder!) It is incumbent upon the RCM team leader to keep RCM in front of management. How best to do this?

Feedback, more specifically feedback in person, is the key to success. Get on management's agenda on a pre-set and continuing basis. Right from the outset, establish a rapport with the member or members of the management team who are ultimately responsible for your project, and convince them that you should present a status report on this very important project at each regular project review meeting. As much as possible, you need to control your destiny and that of the project's. Keep the reporting short and focused on progress towards the goal, not on details. Be as positive as possible—this is a good place to bring up the IOIs which surfaced during the week and what their potential seems to be. All of this is aimed at keeping management's interest and attention; if you are seen as saving and not costing money, you will have a sympathetic ear. Gaining management's awareness will ease the few times when things may have gone astray and you may need their understanding and help to achieve a mid-course correction. Speaking of requesting action from management, be sure to have a suggested solution in mind when you ask for their assistance. You are there to gain their concurrence rather than to ask directly for a solution. Remember, you are dealing with somebody who always reports to somebody else. So give them something to carry up the corporate ladder that speaks well on how RCM is leading the way towards achieving *World Class* distinction and recognition.

We have discussed in the previous paragraph the importance of maintaining an open feedback to management. Just as important are feed-forward comments from management to the troops in the trenches. Management's comments need to be often, positive, and visible. They need to assure those doing RCM that the program is supported and recognized for its contribution to the bottom line and the company's progress towards achieving *World Class* status. The most successful RCM programs are those where management, at all levels from the top to the bottom, makes a concerted effort to have a keen interest, awareness, and presence in the activities being undertaken on the company's behalf.

On two occasions, we encountered management feedback situations that were totally beyond control by the RCM project—situations that ultimately led to failure (i.e., the successful pilot project was abandoned and the entire RCM program was dropped). Both failures had the same root cause—change of key top managers. In both instances, we had originally achieved key management Buy-In and support, and in one instance the pilot project had identified and implemented two IOIs which saved some $300,000 during a major scheduled outage. Basically, the new top managers, who brought in their own people to key plant staff positions, arrived on the scene just as pilot project implementation (Step 8—Task Packaging) was to begin. We instantly lost our hard-won Buy-In and ownership. Despite repeated attempts, we were unable to regain program recognition. With the arrival of the new managers, RCM was a dead issue. It was not their idea, and they had their own agenda to promote. We do not have any useful advice on how to cope with such a situation. Just be aware, however, that this can happen to you should fate dictate management changes in strategic locations in the organization structure. Again, Buy-In and ownership is so important. You can't win without it.

9.8.2 Using Quantitative Reliability Data

You may have noticed that we have not used any quantitative reliability data in the RCM systems analysis process (Secs. 5.2 to 5.8). In particular, we have not directly introduced any quantitative failure rate (λ) or reliability modeling data anywhere in the seven-step evaluation or prioritizing process. This is a very deliberate decision for the following reasons:

1. The ultimate decisions on PM task need and selection occur at the *failure-mode level*. With the current data-reporting systems at operating plants and facilities, there is rarely any credible quantitative reliability data collected at the failure-mode level; what quantitative data is collected is found at the component level, where PM task selections are not made (or should not be made). Thus, usable quantitative reliability history (for example, failure rate) is usually lacking where it might be helpful to the RCM process. This could change in the future, and perhaps should be reconsidered if such occurs.

2. In fact, however, there is no pressing need to introduce quantitative reliability data into the RCM systems analysis process. Realistic evaluations and decisions, from a maintenance point of view, can be made from the qualitative engineering and logic tree information that is systematically developed in the systems analysis process.

3. In addition, without quantitative data, the credibility of the results cannot be questioned on some abstract discussion of "numbers" validity. Only engineering know-how and related judgments are subject to challenge, and these areas can be more readily resolved.

4. Many people simply do not understand quantitative reliability values; thus their absence avoids unnecessary confusion and misunderstanding. (For example, did you read and understand App. B?)

While some RCM practitioners may feel differently about the preceding points, it is the authors' experience that any introduction of quantitative reliability data or models into the RCM process only clouds the PM issue and raises credibility questions that are of no constructive value. Quantitative reliability data is not required in the selection of functions, the conduct of the FMEA, or the ordering of priorities in the LTA. It is useful, however, in decisions on the PM task frequency if the age–reliability relationship is known. In the majority of cases, however, the age–reliability relationship is not known with any degree of precision, even at the component level (see discussion in Sec. 5.9).

9.8.3 Information Traceability and Coding

It is a practical administrative consideration to address the question of information traceability. When RCM is applied to several systems in a plant, we find that the systems analysis information from Steps 4 to 7 tends to pyramid, with the apex representing the system level of definition. Couple this with the possibility that several systems (i.e., pyramids) will eventually become the plant RCM program, and we can rather easily visualize the necessity for some accounting structure for the RCM information. Such an accounting structure will permit not only traceability down through a specific pyramid (i.e., system), but will also develop the structure that leads to the creation of an electronic file (if hard copy reports are not desired) and a computerized database of certain key data for future reference.

There are several ways to establish information coding for an accounting structure. In your particular situation, there may already be an active CMMS which contains coding for the plants in your company, the systems in these plants, and the components in the systems. The primary need for coding, then, resides with the information that is peculiar to the RCM process. A simple way to handle this coding is shown below for a given system of interest:

Functional system:	X
Function:	.XX
Functional failure:	.XX
Component:	.XX
Failure mode:	.XX
Failure cause:	.XX
PM task:	.XX

Thus, for a given plant and system, each piece of RCM information will have a unique 13-digit number for identification and traceability purposes. While this may seem a bit cumbersome at first glance, this is actually not the case when the systems analysis information is committed to a computer for storage and processing.

Furthermore, the value of such an accounting system becomes clearly evident, even with a single complex system, when you find it necessary to retrieve or cross-reference a piece of systems analysis data. With multiple systems, the numbering structure avoids what otherwise might well become an accounting quagmire by providing a unique identification and label for each piece of RCM data.

9.9 KEY FACTORS IN SUCCESS—AND FAILURE

We have attempted in this chapter to condense many of the salient features that we have learned throughout our years of involvement about making RCM a success. The features, if recognized, planned for, and executed can provide a high probability of success; or, if ignored, will likely contribute to the program's failure.

In summary, we present our Key Features—Do's and Don'ts:

- *Do* obtain Buy-In from all levels, especially the craft people and your peers. Everyone wants to succeed, show them in real terms how it can happen and make them all a part of it.
- *Don't* ignore the financial and budget groups. They have more influence than is generally recognized. So establish a strong relationship and gain concurrence on accepted business costs and ROI calculations.
- *Don't* make RCM just another *flavor of the day*. The benefits are real and everyone should see that RCM is supported by the management team and is here to stay.
- *Do* keep the feedback channels to management and those doing RCM open and active. Other important issues arise daily, but establishing and maintaining a presence for RCM is critical to its long-term success.
- *Do* place the "best" craft people on the RCM teams. What you put into RCM determines what you get out.
- *Do* implement RCM's recommendations as soon as possible, even the simplest change. The sooner you begin to implement, the sooner RCM will become ingrained in the daily plant activities and culture. Do not wait for the big push at the end—you may never have that opportunity!
- *Don't* put your head in the sand and assume that all is well; look for the obstacles and the opportunities around the next corner and act accordingly. Plan, plan, plan, and then plan some more!
- *Don't* let an opportunity go by where you could have touted the improvements made by the program. We cannot emphasize enough that positive visibility is a major key to success.
- *Do* assign a caretaker to all IOIs. IOIs are manna from heaven, they come free of charge, and together they contribute unimaginable wealth.

In summary, stay on top of the game, maintain visibility, and never assume that all is well.

10

THE LIVING RCM PROGRAM

The authors' view of World Class Maintenance was described in Sec. 1.5, and was characterized as consisting of five key ingredients. Given that an organization has achieved such a World Class status (a subject of considerable debate in its own right as to just how this would be determined or measured), it is important to understand how to sustain this status over the long term. Our view is that the RCM process is the key ingredient for the achievement of World Class status because it is the most effective way to "focus resources for the best ROI." Our view also included "measure results" as a necessary ingredient. Thus, the ability to sustain World Class status is largely dependent upon our ability to continuously follow-through on these two ingredients. This is done by actively performing Step 9 of the RCM process—The Living RCM Program.

In this chapter, we will describe the simple steps behind performing a truly effective *Living RCM Program.* We will begin by first exploring the need to have such a program, then discuss the factors that drive that need, which will include suggesting some useful metrics to monitor the current health and future benefit of the preventive maintenance program.

10.1 DEFINITION AND NEED

A Living RCM Program is a three-part process conducted continually over time to: (1) validate the preventive maintenance decisions that were made in the RCM baselines; (2) provide for the reassessment of those PM decisions; and (3) make any necessary adjustments to the PM program and the RCM baseline definitions. A Living RCM Program assures continual improvement in the cost-effective operation and maintenance of the plant, but we must also employ some effective metrics to know where the program stands on the above three points.

The RCM processes described in Chapters 5 and 7 are "one-shot" efforts that essentially provide a baseline definition of the PM program for the system in question. However, we need to recognize three technical factors where some continuing RCM program activity is required in order to continuously harvest the full potential of the RCM process:

1. The RCM process is not perfect, and may require periodic adjustments to the baseline results.
2. The plant itself is not a constant since design, equipment, and operating procedures may change over time, and these changes can affect the baseline results.
3. Knowledge grows both in terms of our understanding of how the plant equipment behaves and how new technology can further improve our baseline results.

A fourth and equally important factor is to measure actual versus planned improvements on a continuing basis.

There are other factors, than the four just listed, that influence the perception of maintenance program effectiveness. We will not devote much explanation to these factors, other than to make the reader aware of their existence and warn that they should not be confused with the four listed above. Care must be exercised in distinguishing between those that are a direct consequence of implementing the RCM recommendations and those that are not.

- The failure occurred in a system not included in the RCM program.
- The failure occurred in another system outside the boundary of the RCM system being monitored, resulting in a cascading or secondary failure in an RCM system.
- The failure occurred in an RCM-based system where the RCM PM task recommendations had not yet had time to be effective.
- RCM recommendations have not been implemented.
- The PM task may not have been performed as required.
- Characteristics of the failure mechanism related to age and use were not fully understood—if at all—or the failure was not maintenance preventable, e.g. random failure with a very short time fuse (digital electronics), or the failure resulted from non-maintenance actions (operator error).
- The failure may be the result of maintenance human action of commission or omission.

10.2 THE FOUR FACTORS OF THE LIVING RCM PROGRAM

The continuing activity suggested by the above four factors is what we call "The Living RCM Program." Here, we will briefly discuss the first three factors, and then devote an entire section to the fourth factor—Program Measurements.

10.2.1 Adjustments to the Baseline RCM Analysis Results

The main consideration here revolves around the question of whether the "baseline definition" is totally correct. The most likely answer is "no—not totally." But how do we know this? Back in RCM Steps 5, 6, and 7, we made decisions on what failure modes could reasonably be expected to occur, and what appropriate PM actions should be taken. So if a failure mode occurs that is not directly linked to an RTF decision, we have encountered an Unexpected Failure Mode. And even RTFs with a high rate of occurrence may classify as Unexpected Failure Modes. The best way to watch for this situation is to periodically review the corrective maintenance actions that have been recorded. This measure will provide a direct reading of where the RCM baseline definition either was in error or missed something important. If we are experiencing Unexpected Failure Modes in spite of our PM actions, or a higher than anticipated rate of RTF, then the original PM action selections may be the wrong thing to do and some adjustment to the type of tasks performed (for example, task content and/or frequency) may be in order. If an Unexpected Failure can be traced to a failure mode that is not covered in the baseline definition, then we need to include it as an amendment to the original systems analysis process, and reevaluate the necessity for including a new PM task.

We generally use the term failure to represent a point of equipment degradation where a functional failure has already occurred. If this failure has occurred for any of the above reasons, then it is an Unexpected Failure. However, there are other degraded conditional states where functional failure has not occurred but the anticipated condition of the equipment is not what was expected. Information that is useful in tracking these unexpected conditional states can be divided into three sets of as-found conditions where each set has its own response in The Living RCM program. These as-found conditional states that can be recorded in conjunction with most PM or CM actions are as follows:

1. *Superior*—No discernable degradation on *any* subcomponents.
2. *Satisfactory*—*All* subcomponents show at least the anticipated condition, and are within tolerance; current task and interval appear to be correct.
3. *Unacceptable*—One or more subcomponents are outside tolerance, or action should have been taken before now; may not continue to function properly at current PM interval.

As-Found condition 1—Superior—is a good candidate for Age Exploration where an increase in PM task frequency would most likely be appropriate.

As-Found condition 2—Satisfactory—basically says the PM task content and frequency are correct and nothing should be done to either, just continue the monitoring program. As-Found condition 3—Unacceptable—indicates that a functional failure is imminent, repair/replacement should have already occurred, and that a review of the PM task content along with its interval is required before the next cycle.

10.2.2 Plant Modifications

Even though a plant or facility is literally cast in concrete, it is rare indeed to encounter a situation where there are no changes to the plant over its operating lifetime. These changes occur for a variety of reasons—such as capacity enlargement, productivity improvements, safety and environmental enhancements, regulatory enforcement, and obsolescence. They may involve new additions, redesign of existing systems, replacement of components with upgraded features, and alterations to operating procedures to reduce equipment stress or increase efficiency. Any such change should be reviewed against the RCM-based PM baseline definition to ascertain whether new or modified PM tasks are needed and, in some instances, to delete PM tasks that are no longer applicable and effective.

10.2.3 New Information

Our knowledge base is continually increasing. We learn about the "personality of the plant" as our operating experience grows, and (hopefully) we collect operating and maintenance data which expands our ability to analyze and understand the equipment behavior. This expanded knowledge of the plant behavior may tell us that the PM program requires some adjustments. For example, our knowledge acquired from the Age Exploration process permits us to adjust task intervals. We must also recognize that predictive maintenance technology is expanding. New techniques for the condition-directed tasks are emerging as you read this book. Thus, we may find that PM task effectiveness can be increased with this new knowledge if we use it to our advantage.

10.3 PROGRAM MEASUREMENT

Even if the baseline definition never changes, at a minimum we should measure the benefits derived from the RCM program as a part of the routine plant operating reports to management. Of course, management will be particularly interested in how RCM has impacted the bottom line. Changes to the baseline definition should also be measured to assure that the PM task effectiveness criteria have, in fact, been optimized. Suffice it to say that such measurements can be somewhat difficult to obtain—for example, they can be so global in nature that it becomes very difficult to sort out the parameters that are governing the observed result;

on the other hand, they could be so abstract that it is impossible to clearly define a meaningful message. Plant availability or capacity factor is a typical global measurement; it is a very important measurement, but so many factors can influence its rise or fall that it may be next to impossible to pinpoint the precise reasons for a change. The PM to CM cost ratio is, in the authors' view, a very abstract measurement, and one that has neither good nor bad values. For example, in a well-constructed RCM program, we have seen that RTF decisions are an important part of the total makeup. How does one account for the influence of RTF decisions on the PM:CM cost ratio in deciding what ratio values are good or bad? Given the preceding caveats, there are three measurements that historically have proven to be useful, at a minimum:

1. *Unexpected failures.* As noted previously, this measurement is very valuable in fine-tuning the PM baseline definition for each system. Over time, the occurrence of unexpected failures should approach zero.
2. *Plant availability.* Even though this is a global measure, it does fairly represent a very important indicator of plant performance. And, as plant availability increases, cost avoidance accruals can be a major bottom-line benefit, i.e., avoidance of costs or income losses associated with plant downtime. For pilot and early-stage RCM programs, this measurement may be reduced in focus and reported only on the systems that have received an RCM treatment.
3. *PM + CM costs.* This total cost figure, tracked over time, gives an excellent measure of just how the RCM program is affecting maintenance expenses. It is the total that counts, not the individual values. If RCM is doing its job, this total will decrease over time. If PM and CM costs are reported separately, a very distorted perspective may be sent to management. Consider this—your plant has been reactive for years (lots of CM—not much PM). RCM now introduces a very proactive maintenance program. Initially, therefore, PM costs increase while CM costs stay constant (the beneficial impact of PM has not yet been reflected in the management reports). So, management's reaction is "how come I paid to put more PM into place—but nothing happens to CM—so RCM did nothing but increase my costs?" While there is a lag in PM + CM costs, they can be justified—but individual measurements may be difficult to explain.

There is one area of maintenance cost reporting that continues to cause varying degrees of controversy and concern. We refer here to decisions on how to charge the cost of maintenance actions (usually hands-on types of actions) when they are the result of findings associated with CD and FF preventive maintenance tasks. For example, vibration sensors on certain types of rotating equipment may be set to automatically alarm when displacement values of 0.004 inches are recorded. This alarm is telling us that bearing wear or deterioration has reached a point of incipient failure that will require bearing replacement within the next 45 days. So the replacement is scheduled (hopefully at an opportune time when shutdown of the equipment will not disrupt production) and performed. Now, is this cost chargeable as PM or CM? Our contention is that this is a PM cost simply because the CD task clearly implied (if not explicitly stated) that some form of preventive action, triggered by scheduled monitoring, would likely occur sometime during the equipment operating lifetime. This was clearly a part of the original intent in scheduling the CD task. The same scenario applies equally to FF tasks which are planned to occur at some specified frequency, and also to fix the failure if that is what the finding task discovers. Charging such costs to the CM side of the ledger, to us, distorts the PM versus CM picture.

A slightly more difficult question arises on how to charge costs related to an RTF decision. Recall that RTF is a deliberate decision—i.e., a pre-planned action—to wait until you must act. We consider the repair/replace action associated with RTF to also be a PM cost. But notice, if you follow our suggestion above to always track and report PM + CM costs as a single value, much of this controversy and concern simply disappears.

10.4 Reviewing the Living RCM Program

The final question to consider is the frequency with which The Living Program formal review should be conducted on the baseline definition of each system. Please recall that the accumulation of information for each of the previously discussed four factors is, itself, an ongoing and continuous process. The question here, then, is directed at how often we need to take this information and specifically compare it to the existing RCM program documentation. To a large degree, the answer is strictly a judgment call. In the case of a major unexpected failure, a major plant modification, and the like, an immediate review may be in order. However, it is more likely that the formal review should be conducted every 12 to 24 months. With this interval length, the resources required for The Living Program are fairly minimal, and we have allowed sufficient time for items requiring adjustment to appear. In all likelihood, the need for adjustments will diminish in time, and the frequency of The Living Program reviews will increase to the 36-month range and longer. However, the need to maintain a continual watch on the pulse of the entire PM program cannot be overemphasized.

10.5 Living RCM Program Process

Ensuring that the benefits of an RCM program are fully realized and implemented is the goal of The Living RCM Program. The purpose of any living program is a constant incremental improvement in PM task effectiveness and reduction of the total cost of maintenance.

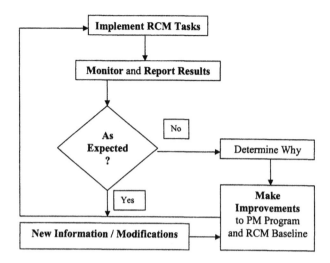

Figure 10.1 Living RCM program—a simplified process diagram.

The process diagram in Figure 10.1 can be used as a fundamental starting point for any user wishing to develop an effective Living RCM Program and can be readily adapted for application by any organization. The process presented is based on Deming's principle of the Plan–Do–Check–Act cycle. It is a simplified version, similar to those employed by many 6 Sigma/TQM adherents in their approaches to World Class Maintenance.

This diagram, while simple, is the core of any good incremental improvement process. Monitoring of results may take many forms and must be structured to the individual organization. The point to remember is that to keep the PM program current and profitable you must <u>first monitor</u> it and <u>then make any necessary adjustments.</u>

11

SOFTWARE SUPPORT

In this chapter we will attempt to de-mystify RCM software. This chapter is not a "how to" treatise meant to enlighten the reader on what keystrokes need to be punched. Instead, we are going to make use of this space to show how beneficial the proper use of a well structured RCM software product can be, both from a time and from a cost perspective, and what is the authors' pick in RCM software and why.

11.1 RCM SOFTWARE—A HISTORICAL JOURNEY

Throughout this book, we have endeavored to demonstrate that the development of a truly effective PM program will be achieved only by intelligently and diligently applying the four basic principles of RCM. As you no doubt have already deduced, software can indeed partially automate the RCM process. However, we should point out that RCM has been performed successfully for years without the aid and benefits of the computer age. But RCM, before software, was by far very labor intensive due to the manual recording of the analysis information. The work required to perform and record the FMEA analysis was tedious enough, but when changes or new failure modes had to be added—well, it simply could be too much. (Manual record keeping has been ascribed even to driving a medieval monk to mutter incessantly and to repeatedly bang his head on his desk.) Now, add the difficulty of extracting data from the analysis sheets to produce useful summaries and reports when your analysis sheets were covered by correction fluid and hopelessly out of sequence. Well, you get the picture. It was not a very efficient one. Enter the computer age.

For some, the prehistoric times of the late 1980s and the early 1990s saw the widespread introduction of the PC, the personal computer, to the business world.

It was use of the PC that drove the need for better and more user-friendly software applications. In the beginning, word processing was all that there was and, while it produced legible reports, it was not suited to the team environment of RCM analyses. A few hearty souls used spreadsheets to record the analysts' decisions—this was a breakthrough. A spreadsheet was easy to read and print out (i.e., make a report) and, with a strong heart and some computer knowledge, could be manipulated to rearrange and keep orderly those changes that drove our medieval monks to madness. The more adventurous saw RCM as a database application, so they developed spreadsheet-like data tables to hold the details of the analysis. While these attempts were useful and on the right track, they were mostly performed by those doing the RCM analysis, so they remained crude, not very user friendly, and most lacked any real sophistication to effectively tie the various steps of the RCM process together. These locally developed applications did not forward the right information and decisions made in earlier analysis steps to the next and subsequent steps. Simply, the analysis could not be easily performed without constantly having to refer back to prior data sheets in order to keep track of what you were doing and why.

All of the preceding problems and needs led the good RCM practitioners to develop their own RCM software packages, and the authors are no different. However, we did have an advantage. Our advantage was a simple, successful, and time-proven RCM process.

Albeit, our RCM process is virtually unchanged from the one developed in the early days of RCM and applied to the Boeing 747-100. But it is that success that makes our preferred software what it is—fast, reliable, and user-friendly. Those qualities will be detailed later in the chapter. We will even reveal our pick for RCM software. (For those of you who cannot wait, turn to Sec. 11.5, but just be sure to read the rest of this chapter to gain a complete view on how software supports RCM.)

11.2 AN OBSERVATION

If you, the reader, will allow a dramatic metaphor, we will begin with an observation. Software, like fire, is a creature comfort that has through the ages (mere decades in the case of software) become a necessity, and, as with fire, we either bask in its warming glow or crouch in fear at the conflagration if it is allowed to get out of control. The trick, as they say, is to be its master and not its pawn. To master such a powerful force, one must understand its strengths and its weaknesses, taking advantage of the former and minimizing the latter.

A word of *caution*—be not led astray by promises of ease, reduced effort, streamlined processes, and the like. The need to complete our projects in the shortest time possible, with the least effort, and at minimal expense can sometimes drive us to take risky shortcuts. These shortcuts seemed prudent in the early going, and if success were measured only in our ability to fill binders with paper, then any old software package would do. To be truly successful, our RCM analysis, when completed, has to be directly translatable and useful to those on the shop floor. Without the support and buy-in of our peers, our bosses, and the craftsmen who will make use of our findings, we only have a pile of paper—just another flavor-of-the-day. If we are to avoid being caught on our own petard, what attributes should our RCM software possess that will make it an ally of the RCM practitioner?

First and primarily, software must support the RCM process. By that we mean that it must adhere to and follow the four principles of RCM we outlined earlier. The somewhat cultist reliance on the computer has brought many maintenance practitioners to a rather risky conclusion and, in the authors' opinion, a potentially dangerous one—that software alone is all that one requires to produce successful PM program improvements. Stated another way, all that is needed is the "tool," which is viewed by some as a means unto itself. If some software application tells us that it is so, the current logic would have us believe that it must be so. The desire is to push a button, copy a file, and to have automatic work completion. But RCM is really a decision process that requires human experience and knowledge to successfully employ. Software must only enhance our RCM process by easing the time and paperwork burden—it cannot and should not do the thinking for us. Ultimately, no matter how automatic, or how many bells and whistles software has, it is the knowledge of the RCM team and the skill of the RCM Facilitator that wins the day, not the software application. To be sure, the use of software as a supportive tool of RCM has shortened the time and reduced the effort of completing the analysis documentation. This reduction in time equates to a direct and substantial reduction in cost. The proper software has indeed made RCM cost-effective!

11.3 Reducing RCM Analysis Effort and Cost

The purpose of RCM software is to reduce in real time, and therefore cost, the level of effort necessary to perform and capture the significant details of an RCM analysis. The right software package can and does perform these beneficial tasks. A well conceived application should, as a minimum, perform with ease, clarity, and speed the Classical RCM approach needed for the 80/20 critical and bad-actor systems. In addition, RCM software should be versatile enough to adapt itself without any loss of ease-of-use, information detail, speed, or any special requirements to handling the 20/80 well-behaved systems, using the Abbreviated Classical RCM™ approach that is defined in Chapter 7.

Just as PM program improvements may reduce the number of times a PM task is to be performed, e.g., by lengthening the time between the scheduled events, reduction in the number of input keystrokes required to complete the RCM analysis likewise has a similar effect—saving time and money. The net effect of simplifying the input and reducing the effort to use and reuse that information is to *speed up* the entire process. This is what software should do.

The tendency here is to complicate the simple and straightforward RCM process by adding software capability, e.g., self-contained pick lists of component types, failure modes and causes, and PM tasks, to name a few. Most of these added capabilities, while seemingly useful, are not expansive enough to provide a general usefulness. For example, they do not contain all the possible credible failure modes, and most definitely do not provide a list of useful functions and functional failures. The presence of too many "pick lists" increases both the complexity and input time requirements of the analysis, simply because selecting from pick lists is generally slower than directly typing what you know, especially if the lists are long and not focused on your particular industry and facility type. Extraordinary software capabilities do not improve upon the RCM process, and they can add substantially to the base cost of the RCM software package as well as that of its actual use.

A far more dangerous concern is that, in making the analysis too automatic and constraining selections to the preset and fixed "pick lists," the inclination is to accept without challenge what is presented, therefore short-circuiting the human decision process. There is no one single cookie-cutter approach to RCM that can predetermine, for any given unique set of equipment and operating profiles, how they should be maintained to truly reflect the long-term goals and need of that company.

The Holy Grail of the preventive maintenance practitioners has always been to discover a simple approach to determine appropriate PM tasks that gives consistent results for the lowest overall investment. Time and again RCM has shown itself to be the solution, but there is always a cost to be paid. RCM requires the expenditure of time, resources, manpower, and of course money. A truly practical RCM software application must be simple, easy to use, one that does not over-complicate the process with unnecessary capabilities, and one that adheres to the principles of RCM. The only automatic decisions that should be made by the software are those that are the natural result or outcome of the RCM process itself, e.g., the analyst's decision on run-to-failure candidates with an automatic decision to submit each candidate to a final sanity check. A discussion on which attributes software should have is presented in the next section.

So how much is a good piece of RCM software worth? Replacing manual recording of the systems analysis process with an efficient and practical software analysis tool has (in the authors' experience) typically reduced efforts by about 20 percent on average. A reduction of this magnitude can be worth $10,000 or more by lessening the time required by the RCM team to complete each system study. For the software that we recommend in Sec. 11.5, a single set can more than pay for itself in its first use!

11.4 USEFUL ATTRIBUTES AND CAPABILITIES

What are some of the attributes and capabilities that RCM software should possess? First and foremost, RCM software should always adhere to the four basic principles of RCM. In our experience, most of the RCM software available today does not meet this first test. Second, the design should provide *user-friendly* forms where the information can be easily entered, and then viewed at a glance to provide understandability to all who see it. Again, we have not found what we would consider to be *user-friendly* forms and procedures in virtually all available RCM software. Third, feed-forward capabilities that eliminate unnecessary re-entering of data while ensuring data consistency and accuracy are another hallmark of a good RCM software application. This is one of the primary features that makes our recommended software user-friendly.

Some valuable and time-saving capabilities are, for example:

- The software should ease the criticality decision process for a component and its failure modes.
- A module where a written system description, the start of a well-documented PM basis, can be entered which contains the system's purpose for existence, the components encompassed within its boundaries, and how those components function in support of the system and the plant as a whole.
- Development of the *Functional Block Diagram* should be facilitated by the direct and easy import of pictures, diagrams, or drawings from many of the popular software packages, and an *in-situ* development of the diagram from within the software itself must be available.
- The software must be capable of developing a list of the system's component tags (IDs) along with their descriptions. A link to the client's internal IT systems could allow for the importing of this information, another potential time saver.
- The component IDs and their descriptions, and the functions and functional failures developed in early steps, are automatically made available in the subsequent steps.
- Failure modes and causes assigned to each component ID, along with that component and its description, are automatically forwarded in a sorted numerical sequence to the appropriate steps.
- All failure modes, along with their Applicable and Effective PM tasks, are conveniently summarized in a task comparison step, where the recommended improvements suggested in the analysis can be compared to the existing PM program.
- Any RCM software worth its salt has to determine and flag a component failure if it could cause personal injury (a safety issue), or one that, should it occur, is hidden from and not apparent to the operating staff.

- Components found to be non-functionally critical should be automatically forwarded to a sanity check where additional considerations, such as cost to repair or regulatory commitments, can be reviewed. This will provide additional information on the need to perform basic maintenance or to assign the component's failure mode to a run-to-failure status.
- There should also be a convenient place for capturing Items Of Interest (IOIs), those *pearls of wisdom* that emerge during the analysis, suggesting new and alternative ways to increase safety, reliability, operability, and maintainability, to name a few.
- Report printing capabilities will allow the user to print the entire analysis (i.e., a presentation report including cover page), or any other combination of pages.
- On larger forms (e.g. FMEA, Task Selection), information contained on the left side of the form would be re-displayed and kept in view of the user as the analysis for the selected component scrolls to the right, thus constantly tying the analysis process to the selected component.

In addition to those already noted, other important and useful efficiencies are:

- The software should contain a Help capability that assists the user in the use of the tool. Help should be of two forms: (1) software usage, and (2) RCM process.
- Each form should contain convenient navigation menus and buttons, allowing the user to efficiently and easily move to any desired analysis step.
- Some (RCM) steps, such as the FMEA and Task Selection, are best presented and printed in an easy-to-read spreadsheet format.
- Analysis data should be exportable, thus allowing a system engineer to access important equipment failure information and the details of the RCM analysis process.

Finally, these supplementary attributes could elevate a good RCM software package, raising it to the extraordinary. While not necessary to the performance of an RCM analysis, they raise the bar and make an analysis tool a complete package. They are:

- Link the RCM analysis software to the client's databases, allowing for the direct importing of data such as component tag numbers, component descriptions, failure mode information, and current tasks and their intervals.
- Include a report module to summarize and group the RCM selected task by component tag or ID, task name, and interval, producing a Task Rollup. Task rollups are useful in the modification of the existing PM program and in the development of new PM tasks. A further enhancement

would be to include the failure modes addressed by the recommended tasks to assist in the development of comprehensive maintenance and operation procedures and instructions.

- Include a module that would provide an ROI (Return-On-Investment) capability, utilizing client cost data. This could demonstrate to management in real and acceptable terms the value of improving the maintenance program.
- Finally and no less important, include a capability to develop and maintain an RCM/PM Living Program that would provide the needed emphasis to keep the new maintenance program current, alive, and returning value to the client. This module should have a link back to the RCM analysis so it too can be maintained and updated as situations change.

While no one RCM software application, at present, encompasses all of the above attributes and capabilities, one does come extremely close. It is the software used by the authors in our RCM work—it is our pick. It lacks only the final four attributes above, and, at this writing, these improvements are in process.

11.5 RCM SOFTWARE—OUR PICK

Our pick—the authors' choice—is the *RCM WorkSaver* by JMS Software of San Jose, California. The *RCM WorkSaver* was specifically designed to follow the Classical RCM approach used by the authors of this book; it can also effortlessly manage the Abbreviated Classical RCM™ Process. The *RCM WorkSaver* can be applied universally to any class of plant, facility, or component type. In fact, the *RCM WorkSaver* has successfully been used in such diverse industries as aircraft manufacture, power generation, paper production, aerospace testing (USAF and NASA), and naval support facilities. It has even been used to develop the PM program for a fuel cell.

The *RCM WorkSaver* was born out of necessity. The authors had long practiced RCM using the previously mentioned and formidable manual approach; we had even tried a few of the client-designed RCM applications. All were found to be lacking in some respects. Some were lacking in adherence to the Classical RCM process, most in ease-of-use, and in the reduction of the time required to complete, print, and present an RCM analysis. The solution, as it turned out, was simple.

We located a software developer who had, in one of their past lives, participated in one of our RCM projects. They knew and understood the RCM process we employed, and they knew how to write software. JMS Software was willing to take on the effort and, more importantly, to tailor the software to the specific process described in Chapter 5. It was a natural fit and has been a successful collaboration since the beginning. JMS Software and the *RCM WorkSaver* can be found on the web at www.jmssoft.com.

In summary, a good RCM software application can do many wonderful things, and we feel that we have one. As a minimum, it should follow the four basic principles of RCM, and it should ease and speed up the overall analysis process. In accomplishing these objectives, RCM software should not drive the analysis but support and enhance the effort. Remember, just because it is computerized, this does not mean it is of any real value unless it possesses the several attributes outlined previously.

12

INDUSTRIAL EXPERIENCE WITH CLASSICAL RCM

The Classical RCM process, described in Chapter 5, has been successfully applied in the industrial arena on scores of systems over the past 20-plus years. The authors have been privileged to personally guide and participate in over 50 of these successful projects. In this chapter, we are especially pleased to present some very specific experience that has been accomplished with Classical RCM.

In Sec. 12.1, we will briefly discuss examples of actual results from the systems analysis process at the component level. These have been selected to illustrate the type of technical benefits that our clients have realized from their RCM program. In Sec. 12.2, we present seven Classical RCM case studies which describe in some detail specific projects that were performed over a broad cross-section of U.S. industry.

12.1 SELECTED COMPONENT PM TASK COMPARISONS

Fifteen specific component PM task comparisons (RCM versus existing), drawn from the results of the Classical RCM process that was performed on 80/20 systems, are shown in Figure 12.1. These fifteen examples help to illustrate the power of RCM in the PM optimization process. They cover the three most significant areas that our clients have experienced in restructuring their PM program using the Classical RCM process:

- PM task addition
- PM task deletion
- PM task redefinition

	System title	Component / failure mode	RCM program RCM-based task	Freq.	Existing program Current task	Freq.	Comment
1.	Air-cooled condenser	Vacuum deaerator condenser / internal fouling	RTF	-	Clean shell and tube sides (TDI)	1 year	System water quality makes fouling implausible.
2.	Electrical	Transformer fan / seized bearing	RTF	-	Replace bearing (TDI)	3 years	Fans are redundant. Spares are available, and bearing condition at 3 years is satisfactory.
3.	All systems	Manual valves / leaks	Visual inspection during walkdown (FF)	Daily	Repacking (TDI)	1 year	Costly to repack, often causes leaks due to repacking errors.
4.	Main generator	Generator / journal bearing fails	Perform oil analysis (CD)	1 month	None	-	Monitoring very cost effective.
5.	Circulating water	Pump / seized coupling	Inspect for wear and lubricate (TDI)	2000 hours	Repack coupling (TDI)	1 year	History of wear and need for more frequent lubrication.
6.	Feedwater	Pump suction strainer / plugged	Record pressure drop across strainer (CD)	1 month	Clean strainer (TDI)	1 month	Infrequent clogging history. Operations can easily monitor pressure drop.
7.	Condensate	Heater drain motor / seized bearing	Monitor bearing temperature and vibration (CD); Perform oil analysis (CD)	3 months; 18 months	Overhaul (TDI)	6 years	Overhaul history void of problems. Monitoring very cost effective.
8.	Electro-hydraulic control	Load unbalance amplifier / out of adjustment – high	Perform calibration (TD)	18 months	Perform calibration (TD) [Does not include load current card]	18 months	Minor modification to current task was very cost effective.
9.	Electrically operated valve	Actuator / gear wear	Inspect gears (TDI); Test grease for wear metals (TDI)	24 months w/AE; 12 months	None	-	Positioning of valves critical to test performance.
10.	Hydraulically operated valve	Barrel seal / oil ring leak	Visually inspect for leakage (TD)	6 months	None	-	Cost effective task with no intrusive concerns.

Figure 12.1 Selected PM task comparison.

	System title	Component / failure mode	RCM program RCM-based task	Freq.	Existing program Current task	Freq.	Comment
11.	Compressed air	6.9 kV Drive motor / insulation degradation	Motor Current Signature Analysis –MCSA (CD) Acoustic monitoring for coronal discharge (CD)	1 month w/AE 3 months	Winding inspection (TDI)	12 months	New tasks are more effective and non-intrusive.
12.	Lube oil	Lube oil heat exchanger / clogged	Track and Trend lube oil temperature (CD)	1 month w/AE	None None	- -	Non-intrusive cost effective task to determine heater cleanliness
13.	Computer driven router	Router bed / worm ways or bearings	Inspect for contaminated grease and excessive wear, replace only if necessary (TDI)	18 months w/AE	Replace way bearings (TDI)	12 months	Very cost effective change both in machine availability and total O&M costs
14.	Process air	Compressor / dirty blades and diffuser	Vibration analysis with trending (CD) Performance Monitoring to track changes in inlet and outlet pressures (CD)	1 month 6 months	Vibration analysis with trending (CD) None	1 month -	Performance Monitoring task is cost effective, non-intrusive, and informs system engineer about the total machine performance and operational parameters
15.	HVAC – Conditioned Air	Activated carbon filter / saturated	Randomly remove one filter from matrix and send to lab for contamination testing (TDI)	12 months	Randomly puncture one filter and remove a sample and send to lab for testing, making sure to seal puncture (TDI)	12 months	Filters are key to maintaining certified air quality. New task does not have a contamination risk as does the current task

Figure 12.1 Continued

PM Task Addition

The most frequent area of restructuring that occurs deals with the addition of PM tasks to handle the potentially critical and/or costly failure modes that were previously not recognized as such. As a rule, these failure modes receive no PM and are tended to only when, by necessity, they must receive corrective maintenance. Such cases are represented by Items #4, 9, 10, and 12 in Figure 12.1. In all four of these examples, the RCM-based PM tasks were addressing the effectiveness issue. That is, they were avoiding costly repairs and/or system and plant downtime.

In some smaller number of cases, a PM task is added to augment an existing PM task in order to increase the effectiveness of the component performance and reliability. This case is represented by Item #14.

PM Task Deletion

The RCM methodology requires, in Feature #4, that every PM task be "Applicable and Effective" (see Sec. 4.4). Failure to meet either criterion means that the task as structured is doing virtually nothing to help prevent the failure mode, or is costing far more than simply repairing the failure mode should it occur. We have found, through experience with our clients over the past 20 years, that 5 percent to 25 percent of the PM tasks in existing (pre-RCM) programs do not pass the "Applicable and Effective" test. In other words, if an organization were to objectively apply this test to their existing PM tasks, sizeable PM resources could be saved, even if nothing else was ever done. (Obviously, we wouldn't recommend that you stop there because this would neglect to act on the really costly issues of corrective maintenance and plant downtime.)

Examples of existing PM tasks that were deleted in favor of doing nothing (i.e., RTF) are represented by Items #1 and 2. Item #1 illustrates an existing task that was "not applicable." In other words, an expensive yearly cleaning exercise was ongoing when in fact, the system had been designed (with attendant costs) to preclude any fouling or clogging in the first place. In Item #2, an expensive replacement action was being done well before it might be needed, but even then redundancy protected the cooling function and spare fans were available if needed. From an "effectiveness" point of view, RTF was the best decision.

PM Task Redefinition

The third area of restructuring, which is essentially an extension of PM task addition, involves a redefinition or perhaps an outright change of the existing PM task. There are several examples of various redefinition cases in Figure 12.1. One type involves changing the existing task to a more suitable (i.e., effective) task that resulted from the RCM process. Examples of this are Items #3, 6, 7, and 11 where an intrusive TDI task is replaced with a non-intrusive CD or FF task. Other cases involve changing the existing task frequency. Item #5 shortened the frequency, and Item #13 extended the frequency. Finally, there are redefinitions that can add incrementally to the existing task to cover another failure mode (Item #8) or can alter the manner in which the task is conducted to reduce risk or cost (Item #15).

Over the past 20 years, the authors have consistently experienced restructuring of existing PM programs by 50 percent or more as a result of implementing the Classical RCM process. In fact, in the case history studies that follow in Sec. 12.2, changes to the existing programs show values consistently in the 50 to 60 percent ranges, with a high of 71 percent. These changes, in turn, have resulted in dramatic reductions in corrective maintenance actions and cost, with attendant gains in product output.

12.2 SELECTED CASE HISTORY STUDIES

In this section, we are very pleased to be able to present the results of seven specific projects that were performed over a broad cross-section of U.S. industry using the Classical RCM process. This has been made possible by the generous consent and assistance that was provided to us by seven outstanding organizations. The seven organizations, represented by four industry sectors, are as follows:

1. *Power generation plants* (Secs. 12.2.1, 12.2.2, 12.2.3)
   - Nuclear—Three Mile Island—Unit 1, AmerGen Energy (prior owner, GPU Nuclear)
   - Fossil (coal)—Neal 4, MidAmerican Energy
   - Fossil (flue gas desulfurization)—Cumberland, Tennessee Valley Authority
2. *Process plant* (Sec. 12.2.4)
   - Bleached market pulp—Leaf River Pulp Operations, Georgia Pacific Corporation
3. *Manufacturing plant* (Sec. 12.2.5)
   - Frederickson Wing Responsibility Center, Boeing Commercial Airplane
4. *Research & Development facilities* (Secs. 12.2.6 and 12.2.7)
   - Arnold Engineering Development Center, USAF/Sverdrup Technology, Inc.
   - Ames Research Center, NASA/Calspan Corporation

The authors wish to express their sincere appreciation for the opportunity to present these informative case history studies.

12.2.1 Three Mile Island—Unit 1 Nuclear Power Plant

Corporate Description

Three Mile Island Nuclear Generating Station was constructed by the original owner, Metropolitan Edison Company, a subsidiary of GPU Utilities. Following the TMI-2 accident (discussed below under "Plant Description"), GPU Nuclear was formed for the sole purpose of operating the Three Mile Island plant and one other plant owned by the company, Oyster Creek Nuclear Generating Station in New Jersey. TMI-1 operated under the ownership of GPU Nuclear until December 1999, when it was purchased by AmerGen Energy, the present owner.

Plant Description

The Three Mile Island nuclear generating station is located on the Susquehanna River about ten miles southeast of Harrisburg, Pennsylvania. Originally, two units (TMI-1 and TMI-2) were constructed and put into operation. In March 1979, TMI-2 experienced the now well-documented core-melt accident due to a "small-break" LOCA (loss of coolant accident) event. As a result of the ensuing investigations and NRC mandates, TMI-1 was shut down for six and one-half years, restarting in October 1985.

TMI-1 is an 870 MWe pressurized water reactor. The nuclear island (i.e., reactor containment and related systems) is a Babcock & Wilcox "once through steam generator" design. TMI-1 began commercial operation in September 1974 and operated at an average capacity factor of 77.2 percent until its shutdown after the TMI-2 accident. After restart, the average capacity factor rose to 83.6 percent and, in 1989, TMI-1 was ranked best in the world from among 359 nuclear plants in 22 nations, with a capacity factor slightly over 100 percent (Ref. 39). TMI-1 continues to be a highly reliable facility and has had a number of world-record operating cycles. Since 1989, average capacity factor is 91.4 percent.

Some Pertinent Background

The original method used to specify PM tasks at TMI-1 was basically to employ vendor recommendations and test them for reasonableness tempered by experience and judgment. The resulting PM program was subject to continuous review to ascertain correct priorities for resource commitment. The program that evolved was component (not failure mode) oriented and tasks were based on the vendor's input or, in many cases, simply what actions could be done. The resulting task basis was almost exclusively time-directed (TD) tasks—overhauls, calibrations, and various intrusive inspections. When the backlog became too large, arbitrary decisions were frequently made to skip PM tasks if the component was not safety-related or known to be a plant trip initiator. As a result, several important components were left vulnerable to failure (e.g., those providing balance of plant operation and permissive/interlock functions). This traditional approach did not guarantee that all critical components were addressed—that is, until a rash of corrective maintenance (reactionary) actions was experienced.

After the TMI-2 accident, there was a 6½ year hiatus before TMI-1 was restarted in October 1985. During this period, there was a concerted effort to review the entire PM program. This included a detailed revisit to the operating history of both Units 1 and 2, a review of nuclear industry pressurized water reactor maintenance and outage data, and a re-evaluation of the latest vendor recommendations. This process defined what was believed to be a PM program that was among the best in the world.

Plant management had considered the RCM approach during this downtime period, and decided in 1987 to begin a comprehensive RCM effort to independently validate the traditional PM tasks that were in place. As one might suspect, we found that several critical systems were validated and received only minor modifications, but others were found to be lacking in several respects and received major modifications. These changes were a key factor in the increasing capacity factors that were measured in the late 1980s and '90s.

The RCM program that we pursued was initiated in September 1988 and completed in June 1994. The sections that follow below describe that program and some of the results and benefits that it produced.

System Selection

From the outset, the objective was to identify the critical plant systems, and then schedule a program to use the Classical RCM process on each one. Our initial approach to system selection was a modified Delphi process (i.e., a structured opinion poll) which tended to make the choices heavily weighted toward safety considerations rather than maintenance optimization factors that focused on reduction of corrective maintenance and forced outages. (The TMI-2 accident issue was naturally uppermost in our minds.) Two results somewhat surprised us: first, the list was composed almost entirely of safety-related systems; and second, virtually all of these systems had a record of low maintenance costs and very high availability. This, of course, was not where our RCM program needed to focus.

Our consultant (an author of this book) had advised us against our Delphi survey, and recommended that we seek out the 80/20 systems using a Pareto diagram with either corrective maintenance or outage histories for the 100-plus plant systems. (Figure 12.2 is a simplified schematic of the Unit 1 plant and systems.) We re-did the selection process with a very different result. As shown in Figure 12.3, we settled on 28 systems that fit the 80/20 rule, only four of which are safety related (other nuclear plant programs later found the same general result).

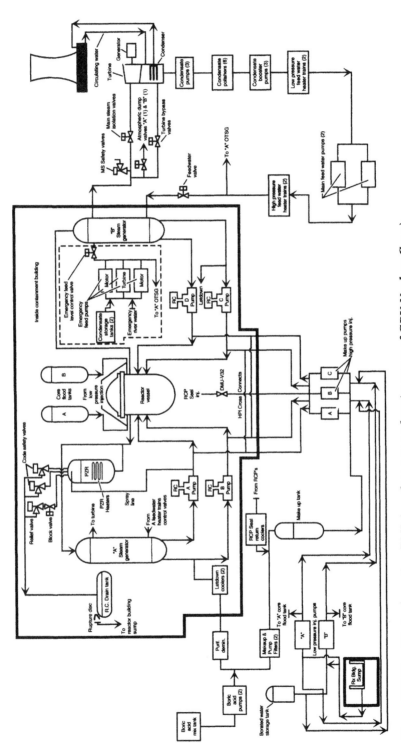

Figure 12.2 Simplified schematic—TMI-1 nuclear power plant (courtesy of GPU Nuclear Corp.).

Condensate
Condensate Polishing (Powdex)
Main Stream
Makeup and Purification (including HPI)
Main Turbine
Turbine Auxiliaries
Main Generator and Auxiliaries
Main Feedwater
Instrument Air
AC and DC Vital Power
Main and Auxiliary Transformers
Decay Heat Removal (including LPI)
Decay Closed Cooling Water
Decay River Water
Nuclear Services Closed Cooling Water
Nuclear Services River Water
Intermediate Closed Cooling Water
Containment Isolation
Reactor Coolant System
Extraction Steam
Heater Drains
Secondary Closed Cooling Water
Circulating Water
Control Rod Water
Emergency Feedwater
Emergency Diesel Generator
Reactor Building Emergency Cooling
Building Spray
Table 1

Figure 12.3 Systems selected for the RCM program.

The first two systems to be evaluated were Main Feedwater (MFW) and Instrument Air (IAS), and we will use results from these two systems to describe the RCM process in the paragraphs below.

Overall, the RCM program ran for almost six years, and completed all 28 systems identified by the 80/20 rule.

RCM Analysis and Results

As noted above, we employed the Classical RCM process (described in Chapter 5) to analyze the 80/20 systems because it was considered important to follow the proven concepts and implementation steps that are so successfully employed in the commercial aviation industry.

Organizationally, a separate RCM function was established in the Maintenance Department with three full-time-equivalent engineers and/or technicians assigned to the RCM team. Several rotating assignments were made on the team with other plant organizations to gain a broad base of exposure to the RCM process. Two important "lessons learned" emerged from our organizational approach: first, all team members were always drawn from the Unit 1 staff (i.e., people "inside the fence" with direct responsibility for running TMI-1 on a daily basis); and second, Operations personnel were always an integral part of the team.

On average, the cost of analysis and implementation was $30,000 per system (in then current year dollars). However, we experienced a learning curve, and later systems were approximately 40 percent less costly to analyze and implement than the first few.

By way of illustration, a review of the Main Feedwater and Instrument Air Systems will be presented below. Figure 12.4 lists the RCM Systems Analysis Profiles, Figure 12.5 the RCM Task Type Profiles, and Figure 12.6 the RCM Task Similarity Profiles.

The Systems Analysis Profile (Figure 12.4) indicates the extent of the evaluations that were performed on each system. Over 1200 failure modes were investigated between the two systems, resulting in 465 separate decisions on what PM actions were necessary. These actions are further summarized in Figures 12.5 and 12.6,

	Main Feedwater	Instrument Air
Functional Subsystems	3	5
Subsystem Functions	69	136
Subsystem Functional Failures	145	187
Failure Modes Analyzed	806	433
RCM-Based PM Tasks	230	235

Figure 12.4 RCM systems analysis profiles.

	Main Feedwater		Instrument Air	
	RCM	Current	RCM	Current
Time Directed (TD)	188	162	169	103
Condition Directed (CD)	18	16	23	9
Failure Finding (FF)	19	18	33	21
Run to Failure (RTF)	5	-	10	-
None	-	34	-	102
Total	230	230	235	235
(RCM Δ)	(+29)		(+92)	

Figure 12.5 RCM task type profiles.

	Main Feedwater	Instrument Air
RCM Task Equals Current Task	175	65
RCM Task Equals Modified Current Task	9	54
RCM Task Recommended, No Current Task Exists	34	102
Current Task Exists, No RCM Task Recommended	12	14
Total	230	235

Figure 12.6 RCM task similarity profiles.

and indicate two significant observations:

1. As noted previously, the RCM process validated that some systems were well structured and recommended only minor PM adjustments—Main Feedwater is one such example. Other systems were found to be lacking in their PM content and significant improvements were identified—Instrument Air proved to be such an example. These differences are most obvious in Figure 12.6 for the statistics reflecting "RCM Task Recommended, No Current Task Exists."

2. Certain systems required a significant increase in PM actions (as noted in #1 above), but much of this included the addition of more non-intrusive Condition-Directed (CD) and Failure-Finding (FF) tasks. Again, Instrument Air is one such example, as indicated on Figure 12.5.

By way of illustration, Figure 12.7 is a "vertical slice" from the IAS which traces the origins and resulting PM actions of a single failure mode through the rigors of Steps 3 to 7 of the Systems Analysis Process. This particular example has three interesting points:

1. The failure mode (desiccant exhausted) is hidden (Step 6: Evident = No). The operator would not normally realize that this had occurred without some "extra" information supplied.

2. The criticality is low (D/C), but eventual water damage in the system would develop.

3. There were no PM tasks in place when this analysis was done. From a cost viewpoint, two simple applicable PM tasks were possible, with the annual inspection scheduled for deletion when the dewpoint alarm was proven sufficient.

Overall, in the 28 systems evaluated, there were 3778 components involved. These components had 4874 existing (pre-RCM) PM tasks, and had 5406 RCM-based PM tasks when the RCM program was completed.

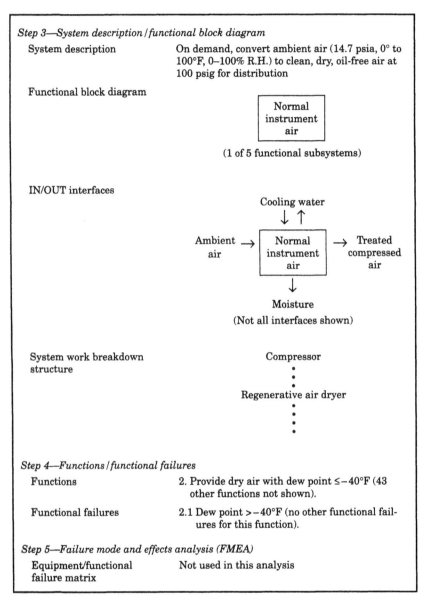

Step 3—System description / functional block diagram

System description   On demand, convert ambient air (14.7 psia, 0° to 100°F, 0–100% R.H.) to clean, dry, oil-free air at 100 psig for distribution

Functional block diagram

Normal
instrument
air

(1 of 5 functional subsystems)

IN/OUT interfaces

Cooling water
↓ ↑

Ambient → | Normal | → Treated
air | instrument | compressed
          | air | air
↓
Moisture
(Not all interfaces shown)

System work breakdown   Compressor
structure   ⋮
                         Regenerative air dryer
                                   ⋮

Step 4—Functions / functional failures

Functions   2. Provide dry air with dew point ≤ −40°F (43 other functions not shown).

Functional failures   2.1 Dew point > −40°F (no other functional failures for this function).

Step 5—Failure mode and effects analysis (FMEA)

Equipment/functional   Not used in this analysis
failure matrix

Figure 12.7 "Vertical Slice"—Instrument air system (GPU Nuclear—Three Mile Island-1).

FMEA						
	Failure	Failure	Failure effects			
Equipment	mode	cause	Local	System	Plant	LTA
IA-Q-Z Air dryer	Desiccant exhausted	Normal wear	Wet desiccant	Wet air discharged into system. Long-term damage accrual.	Degraded operation due to water intrusion into equipment.	Y

Step 6—Logic tree analysis

Failure Mode = Desiccant exhausted

1. Evident: No
2. Safety: No
3. Outage: No

Category D/C
(minor economic problem)

Step 7—Task selection

Applicable task candidates

1. Monitor dew point continuously and alarm at –30°F (CD)
2. Inspect condition and color annually (TD)
3. RTF

Effectiveness factors

Both No. 1 and No. 2 are considered to be cost effective in order to avoid water damage to equipment. RTF not considered to be cost effective.

RCM decision

Perform both continuous dew point monitoring with alarm at –30°F, and the annual inspection. Plan to delete the annual inspection after operational verification of the dew point monitoring has been achieved.

Current task

New equipment. No previous PM tasks were in place.

Figure 12.7 Continued

RCM Implementation

We were aware of various reported difficulties elsewhere in the implementation of RCM recommendations, and structured the following implementation strategy to avoid these problems.

1. Responsibility for RCM analysis <u>and</u> implementation execution was carried within the <u>plant</u> maintenance organization. In fact, members of the analysis team participated directly in the implementation process,

thus providing a direct carryover of the analysis logic into the recommended actions (i.e., buy-in).

2. Management support was visibly demonstrated by imposing the administrative goal to put RCM tasks in place within three months of the issuance of an RCM analysis unless long-term efforts were involved (e.g., new training, new tools, or instrument purchases).

3. The RCM-based PM tasks, including their criticality code, replaced existing tasks in our Generation Management System 2 (GMS 2) and were so labeled as RCM to facilitate tracking and measurement in the GMS 2 Management Action Control (MAC) module.

4. Proposed PM task changes, additions, or deletions were approved by the system engineer, PM foreman for the affected disciplines, and plant operations personnel before they were loaded into GMS 2.

5. If long-term efforts were involved in implementation (e.g., a new task), the RCM analyst directly performed or coordinated the necessary actions and used the MAC module to track progress against schedule.

6. We classified each RCM recommendation into one of six categories to facilitate the review and action process, as follows:
   • Continue existing task
   • Modify existing task
   • Add new task
   • Delete existing task
   • Change task frequency
   • Modify plant

Fortunately, we did have a mature process in place to make procedure and configuration changes as required.

We suggest that success and ease of implementation hinge on a process that incorporates the above six features.

Benefits Achieved

We have previously quoted (in Plant Description) one of the most significant benefits derived from the RCM program—namely, the dramatic rise in the plant capacity factor. While difficult to precisely quantify, we believe that a major share of that increase developed from the progressive implementation of RCM on the 28 identified 80/20 systems, and from the peripheral benefit realized from the discipline and lessons learned from the RCM program.

More specifically, significant reductions in equipment failures and corrective maintenance actions were realized as shown in Figure 12.8. The failure trend at the plant level is decreasing from a high of 950 in the fourth quarter of 1990 to a low of about 600 in the middle of 1993, or a decrease of 37 percent. The flat spots

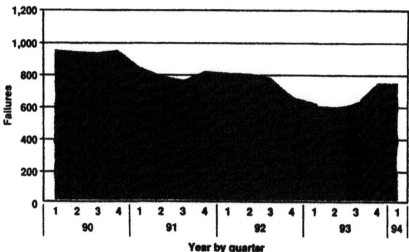

Figure 12.8 Plant level equipment failure trend.

on Figure 12.8 represent time periods affected by refueling outages which occurred in January 1990, October 1991, and September 1993. Note the decreasing trend of the outage flat spots from 950 to 821 to 745 or a reduction of 22 percent. In other words, we were slowly learning that it was not necessary to arbitrarily "maintain" equipment just because it was available during a planned shutdown which actually was triggering failures due to human error in our intrusive (and unnecessary) PM tasks.

In the area of PM task intervals, we introduced a program of Age Exploration on the critical failure modes (Categories A and B), but were reluctant to invoke these changes without more historical backup. However, with the non-critical Category C failure modes, we retained PM tasks when the cost implications so warranted. But here, we did immediately extend task intervals on a number of components, with the resulting annual cost savings shown in Figure 12.9.

Finally, we realized several important qualitative benefits, including the following. We:

- Upgraded spare parts inventory.
- Identified hidden failure modes.
- Discovered previously unknown failure scenarios (note: a probabilistic risk assessment or PRA had been previously performed on TMI-1).

Craft	Man-hour savings/year	Dollars/year
Electrical	2210	$77,350
Instrumentation & Control	2188	76,580
Mechanical	1552	54,320
Utility	1719	60,165*
TOTAL	7669	$268,415

* Assumes $25 per man-hour cost plus an estimated materials and supplies cost of 40% of the man-hour cost.

Figure 12.9 Category savings via interval extension.

- Provided training opportunities for the system engineers and operators (primarily via the FMEA).
- Identified candidates for design enhancements.

In conclusion, we feel that the movement toward "streamlined RCM" may not be a wise decision when dealing with the 80/20 systems. The full benefits are best achieved with the Classical RCM process. Our experience says that the breakeven payback is rather quick, and thus the thorough classical process is more than worth the relatively minor additional effort that may be required.

Acknowledgements

The authors wish to express their appreciation for the efforts put forth by the RCM team and supporting personnel who made this case study possible:

RCM team	*Support*
Barry Fox	Bob Bernard
Dave Bush	Gordon Lawrence
John Pearce	Harold Wilson
Pete Snyder	
Mac Smith (facilitator)	

12.2.2 MidAmerican Energy—Power Plant Coal Processing

Corporate Description

MidAmerican Energy is Iowa's largest utility company, providing an electric and natural gas service to 1.7 million customers in 550 communities in Iowa, Illinois, Nebraska, and South Dakota. MidAmerican Energy was formed by the merger of Midwest Resources and Iowa–Illinois Gas and Electric in 1995. Midwest Resources was formed by the merger of Iowa Resources and Iowa Public Service Company.

MidAmerican Energy operates ten coal-fired power plants in Iowa. They are part owners in another Iowa coal-fired plant and two nuclear plants (one in Illinois and one in Nebraska).

Neal 4 Generating Station

Neal 4 is a 644 MW jointly-owned coal-fired plant operated by MidAmerican and located near Sioux City, Iowa. It began commercial operation in 1979 and generates 4,000,000 MW hrs/yr while burning 2,500,000 tons of Powder River Basin coal. The 1994 UDI Production Costs Report shows Neal 4 as the second lowest-cost plant in the nation ranked by average expenses (including fuel) per net megawatt-hour. The equivalent availability factor and equivalent forced outage rates for Neal 4 from 1990 to 1994 were 84 percent and 5 percent, respectively, while the NERC GADS averages for similar units were 79 and 8.7 percent.

System Selection

Neal 4 consists of some 30 major systems. At the outset, a Pareto diagram based on maintenance costs for a 2-year period was developed in order to define the plant's 80/20 systems. The Coal Processing System, at the top of the list, was selected for an RCM pilot project. This system starts at the exit gate to the plant in the coal silo and terminates at the boiler control inlet gate where the pulverized coal–air mixture is blown into the boiler. There are seven Coal Processing Systems in the plant.

A Functional Block Diagram of the Coal Processing System is shown in Figure 12.10. This system is composed of four distinct subsystems. The complexities in this system mainly reside in the Coal Feed Subsystem and the Pulverizer Subsystem, which contribute the vast majority of the maintenance costs and downtime attributed to the Coal Processing System. The decision was made to select the Coal Feed Subsystem for the pilot RCM project.

The Coal Feed Subsystem

The Coal Feed Subsystem is one of four subsystems in the Coal Processing System. It regulates the rate of coal feed to the pulverizers. The coal feeder is a gravimetric

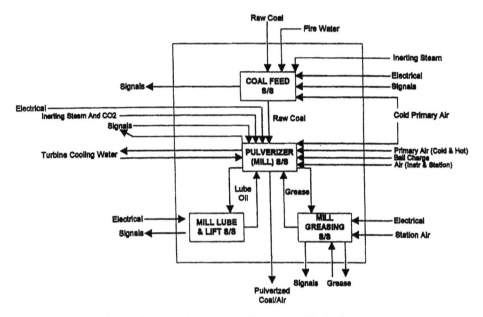

Figure 12.10 Neal 4 coal processing system, functional block diagram.

type which automatically adjusts coal feed for variations in coal density. The coal feed is regulated in response to signals from the pulverizer coal level controller and the combustion control system. One revolution of the belt conveyor head pulley delivers 100 pounds of coal, regardless of density. The weigh-sensing and correction system accounts for changes in coal density by adjusting the coal leveling bar to maintain the 100 pounds per each pulley revolution.

The feeder is equipped with SECOAL double nuclear monitors that sense the presence of coal in the raw coal conduits. The top monitor, located just below the silo exit, will automatically activate vibrators on the side of the silo if it senses a loss of raw coal flow. The bottom monitor, located some five feet above the feeder, will trip the coal feeder if it senses a loss of raw coal flow.

There are two coal feeders per mill, for a total of 14 feeders in the plant. A feeder seal air local-manual control damper is provided to establish seal air in the feeder that will maintain positive pressure with respect to the pulverizer. A weigh chamber seal air shutoff valve also provides seal air to pressurize the weigh chamber. This system also has a manually operated shutoff gate between the mill and the crusher dryer chute. Below the feeder belt is a cleanout conveyor to remove accumulated dust.

- Number of S/S Functions   4
- Number of S/S Functional Failures   8
- Number of components in S/S boundary   14
- Number of failure modes analyzed   130
  - number of hidden failure modes   71 (55%)
- Number of critical failure modes   78 (60%)
  (i.e. A, D/A, B, D/B)
- Number of PM tasks specified   152
  (including run-to-failure decisions)

Figure 12.11 RCM systems analysis profile.

RCM Analysis and Results

The Coal Feed Subsystem employed the 7-step Classical RCM process described in Chapter 5. The data from the analysis process was recorded manually by the team facilitator onto the RCM forms (as the "RCM WorkSaver" software had not yet been developed).

Statistical Summaries

In Figure 12.11, a profile of the analysis process itself is shown, where a total of 152 separate PM task decisions were developed from the 130 failure modes that were identified in each of the 14 Coal Feed Subsystem equipments. These failure modes were derived directly in response to the four functions and eight functional failures defined for the subsystem. Notice that over half of the failure modes are hidden (i.e., the operators would not be aware that something was wrong until an undesirable consequence occurred), a somewhat surprising result in light of the operational experience at Neal 4. We also found that the majority (60%) of those failure modes were critical to either personnel safety or plant outage considerations, thus emphasizing the need for focused PM activity to eliminate such concerns.

In Figure 12.12, the PM Task Type Profile, we compare the current PM tasks with the RCM tasks developed in Step 7 of the analysis. This comparison is quite revealing on two particular points: (1) the total number of <u>active</u> PM tasks nearly tripled (from 34 to 90); and (2) RCM has introduced a significant content of non-intrusive condition-directed and failure-finding tasks (55 versus 12). Both of these points are in direct response to the high content of hidden failures and critical failure modes that we saw in Figure 12.11. The current PM tasks did not address 118 specific items where RCM task decisions were made, and of these

Task Type	RCM	Current
• Time directed		
- Intrusive (TDI)	29	20
- Non-intrusive (TD)	6	2
• Condition directed (CD)	22	7
• Failure finding (FF)	33	5
• Run-to-failure (RTF)	62	--
• None specified	---	118
	152	152

Note: 55 of the "RTFs" were also "NONES"

Figure 12.12 PM task type profile (for failure modes).

RCM retained 55 as deliberate run-to-failure decisions. The other 63 resulted in some form of TD, CD, or FF task (see Figure 12.13, Item 3A).

Another way to examine the RCM versus current PM structure is shown in Figure 12.13, the PM Task Similarity Profile. This comparison reflects where the RCM-based PM tasks either agree with or differ from the current PM tasks. The items denoted by the check mark ($\sqrt{}$) indicate where the RCM process had its greatest impact on optimizing PM resources. In this case, that impact clearly occurred with Item 3A where RCM identified a need for PM action in 63 areas where the current program did nothing. Overall, RCM results introduced a change to 60 percent of the current PM program for the Coal Feed Subsystem.

Selected Task Comparisons

Figure 12.14 illustrates six comparisons that help to demonstrate the kind of RCM analysis results that can impact a PM program that has developed via conventional means over a period of years.

Item 1. This is typical of a finding that frequently develops in the RCM process. A close examination of the continued need to calibrate the Weigh Control Chambers in the feeders led the RCM team to challenge the need for these coal weight measurements at all. A separate evaluation by other plant personnel confirmed that the required coal measurements were taken in two other places, thus negating any need to continue these expensive calibration procedures. In other words, the current task in this instance could not meet the "effective" criteria required by RCM principle #4. The calibrations were deleted and the Weigh Control Chamber was locked out at an annual saving of $50,000.

Item 2. This item illustrates both the application of new PdM technology (thermographics) and the dramatic change in frequency that was developed for all

Similarity Descriptor	Number		Percent
1. RCM = Current (Tasks are identical)	6		4%
2. RCM = Modified Current (Same general task approach)	20	√	13%
3A. RCM Specifies Task – No Current Task Exists (current missed important failure modes)	63	√	41%
3B. RCM Specifies RTF - No Current Task Exists (Similarity probably accidental in most cases)	55		36%
4. RCM Specifies RTF - Current Task Exists (Current approach not cost effective)	7	√	5%
5A. Current Task Exists - No Failure Mode in RCM Analysis	0	√	0%
5B. Current Task Exists RCM Specifies Entirely Different Task	1	√	1%
	152		100%
		√=60%	

Figure 12.13 PM task similarity profile (for failure modes).

greasing tasks in the current program. In essence, greasing was being grossly overdone, resulting in unnecessary costs and in some instances leading to incipient failure conditions due to packed and hardened grease in several bearings and gear boxes.

Item 3. The vibrators located at the interface between the coal silos and the inlet piping to the feeders tend to see little service in the non-winter months when coal freezing/packing jams are absent. Then the winter months hit and suddenly one or more vibrators won't operate. A simple failure-finding task was specified to mitigate this problem. Also, a design modification was introduced to rewire the vibrators for separate test operation (there are two per silo and the noise of operation masked whether both were, in fact, operating during this test).

Item 4. Steam inerting is employed in the Coal Feed Subsystem during a shutdown to preclude the possibility of fire (Neal 4 uses Powder River Basin coal which tends to self-ignite if left standing in the feeder). But, once the shutdown is accomplished,

Component & Failure Mode	RCM Task	Current Task	Comment
1. Weigh Control - all failure modes	RTF	Calibrate (TDI) - 3 months	An equivalent method for weighing coal can be used (7A and 7B scales). There is no negative plant consequence if failure does occur.
2. Feeder Head Pulley -outboard bearing fails/seizes	1. Periodically grease bearings (TDI) - 6 months with AE (AE=Age Exploration) 2. Do thermographic test and trend result - 1 month with AE (use these results to also adjust the greasing interval).	Grease all fittings (TDI) - 1 week --	Generally, all intervals for greasing have been extended from weekly to 3, 6, or 12 months. An example of new technology being utilized in the PM program.
3. Vibrator - coil burned out	1. Periodically operate (FF) - 1 month, 4/1-10/31 - 1 week, 11/1-3/31 2. Modify wiring to run each vibrator in-dividually. (DM)	None Currently, both vibrators operate simultaneously.	Recently did a random check on one, found it did not operate. Then, checked others and found two more that did not operate. Noise makes it virtually impossible to distinguish if one or two vibrators are operating.
4. Feeder Inerting Valve - isolation valve jammed open	Periodically operate to assure it can work (FF) - 12 months (before boiler outage)	None	Team checked these valves and found three that were 'frozen' in the open position.
5. Feeder Control Cabinet - TD-2 timer fails.	RTF *BUT* Operators will verify by inspection that there is no coal on the belt immediately after each shutdown.	None This is not currently done.	This is a new procedure for the operators. This particular failure mode is hidden, and would result in coal left on the belt.
6. Feeder Control Cabinet - MSC relay fails to close	Inspect for drag chain failure (FF). - each shift	None	This inspection must be done by looking directly at the drag chain – not the coupling which could be turning with the worm gear stripped. This latter situation has actually occurred.

Figure 12.14 Other selected results.

the isolation valves are closed while work is performed. If these valves have not been used for some time, they can jam in the normally open position. Again, a simple periodic operation of the valves is done to assure their operability. In a sample check by the RCM team, three valves were found jammed in the open position since the current PM program did not contain this failure-finding task.

Item 5. In a normal feeder shutdown operation, a timer in the control system allows the belt to run for 120 seconds to assure that all coal has been emptied into the pulverizer, thus precluding a potential fire hazard in the feeder. If this timer fails and the belt thus stops fully loaded with coal, no one would normally be aware of the situation unless a fire developed (or they accidentally noticed it). Thus, an Operations SOP was instituted to provide a visual check of the feeder belt status immediately after each shutdown.

Item 6. This failure-finding task was formalized into the Work Management Information System, but most importantly was modified in its procedure to require that the inspection of the drag chain itself be accomplished rather than just a visual check of the coupling between the motor and gearbox located external to the feeder. This is a good example of how some of the Neal 4 equipment history was used to influence the selection and definition of the RCM tasks in Step 7.

Selected IOI Findings

The Classical RCM Process entails a very comprehensive review and evaluation of the selected systems. As a result, the RCM team found a variety of items, non-maintenance related, which deserved further evaluation, correction, and/or action. These items were recorded on a list called "Items of Interest." A few selected examples of these IOIs follow:

1. The flow switches to detect lube oil supply to the mill trunnion bearings had been removed, leaving no lubrication protection to these bearings.
2. The group discovered a method to extinguish a mill drum fire that was not being utilized.
3. A handwheel on the feeder inlet gate was improperly labeled, showing open when it was closed.
4. No one knew what would happen if the SECOAL nuclear coal detector was put to Calibrate position when the feeder was on; would it trip the feeder and present a potentially dangerous condition or not?
5. The SECOAL nuclear coal detector Trouble Alarm on the BTG board actually meant something other than "No Coal to Feeder," as most operators thought.
6. "As found" conditions of corrective and preventive maintenance actions were not being recorded.

In all, a total of 24 IOIs were listed on the Coal Feed Subsystem.

RCM Implementation Process

AMS Associates was retained as a Consultant to guide and facilitate the pilot project. The RCM team for the project was composed of three highly experienced craft personnel—one each from mechanical and electrical/I&C maintenance, and one senior plant operator. As our Consultant noted, use of plant craft personnel for an RCM team has produced some of the most successful RCM programs because of both the extensive knowledge of the plant that they bring to the table and an ownership of the process by the very people who are now expected to execute the results.

The pilot project took 35 days to complete, including the development of specific inputs to the plant Work Management Information System (WMIS) for planning and execution of the task recommendations from Step 7. These 35 days were staggered in one-week increments over a seven-month period in order to preclude any major conflicts or discontinuities in the team's regular work assignments. In the future, the schedules for RCM projects may be compressed somewhat closer; but, in general, we find that some form of staggered one-week intervals works very well for a team composed of craft personnel. With the training experience now successfully achieved, we expect that future projects on 80/20 systems will be done, on average, in 25 days. Use of the "RCM WorkSaver" software should further reduce this to about 20 days.

A smooth transition from the analysis findings in Step 7 to the initiation of the RCM-based PM tasks on the floor was the result of five key actions:

1. The RCM team presented an explanation of the RCM methodology and the analysis results to an all-hands meeting—i.e., craft technicians, supervisors, and managers. Thus, there were no surprises or hidden agendas, and the initial phase of plant-wide buy-in was achieved <u>before</u> any PM task changes were made.
2. The analysis results were thoroughly reviewed and approved by the plant manager and his staff <u>before</u> any changes in the PM task orders were made.
3. The Analysis Team prepared expanded definitions of the Step 7 PM task results to accommodate their accurate entry into the WMIS (see Figure 12.15).
4. All current PM tasks (except the six retained where RCM and current tasks were identical) were purged from the WMIS to prevent any confusing overlaps.

WMIS INPUT FORM

Task Title: *Ultrasound*

Requested By: RCM # *11- 015*

Craft: *M*

Equipment: *Crusher Dryer Outlet Gate*   **Equipment #:**

Work Task:

Take Ultrasound (3 Meter) readings and trend. Initially, take baseline readings on all (6.1) 14 Gates - then select four of the 14 Gates (at random) for continued readings at the frequency shown below. The selected four Gates should be from those used the most.

Frequency: *6 month with A.E.*

Crew Size/MH (Tot): *One / 6 hours (for all 14 Gates) One / 2 hours (for 4 Gates)*

Special Instructions:

Sufficient readings must be taken to assure wall thickness integrity in the circumferential region between the bottom flange and 6 inches above the bottom flange.

Additional Information Required To Implement Work Task:

1. A definition of "acceptable wall-thickness" must be established.

2. A consistent pattern for the readings must be defined for use in establishing trend information (e.g. at 90° mid-point locations for use in trending).

Entered By: *RCM Team 2/22/96* **Reviewed By:**

Approved By:

Figure 12.15 WMIS input form.

5. Operations personnel agreed to accept formal responsibility for a significant role in the conduct of several CD and FF tasks—a role that had previously been non-existent or only informally conducted.

Benefits Realized

The Neal 4 RCM Pilot Project was conducted with two objectives in mind: (1) to demonstrate the benefits available to Neal 4 from the Classical RCM process; and (2) to train a team of plant personnel in exactly how to perform the 7-step systems analysis process. We have successfully achieved both objectives.

In terms of economic benefits achieved from an RCM project, we believe that the savings in PM man-hours, overall, will probably be small. In fact, this specific pilot project almost doubled the RCM task man-hours with respect to the current PM task man-hours. This increase, however, is somewhat artificially inflated because we are formalizing some of the operator PM tasks in our CMMS which heretofore were done only informally and on a random basis. We expect that the major economic benefits will accrue in three areas:

1. Corrective maintenance (CM) costs will be dramatically reduced owing to the effectiveness that results from the detailed RCM process. We estimate reductions in the 40 to 60 percent range, or better.
2. The reduced CM activity will naturally be of benefit in reducing the forced outage rate. It is difficult to estimate what this might be (Neal 4 already enjoys an excellent EFOR history), but even small reductions here translate into meaningful O&M cost reductions.
3. The tangential benefits realized from this Classical RCM process may well turn out to be a very major factor in the economic picture. While hard to estimate, the pilot project Items Of Interest (described above) indicate that these added findings could equal or exceed the maintenance savings on some systems. In fact, these added findings are one of the "surprise" results that have made us feel so very positive about the Classical RCM approach. Without such a comprehensive analysis process, we fail to see how this could occur to any great extent.

Acknowledgements

The authors wish to express their appreciation for the efforts put forth by the RCM team and supporting personnel who made this case study possible:

RCM team	*Support*
Jeff Delzell	Rod Hefner
Joe Pithan	Dana Ralston
John Riker	Chuck Spooner
Mac Smith (facilitator)	

12.2.3 Tennessee Valley Authority (TVA)—Cumberland Plant FGD

Corporate and Plant Description

TVA is a wholly owned corporate agency of the United States which is not only a regional economic development agency but also the nation's largest public power system. The power and river management portions of TVA are not subsidized by the government; they are paid primarily through sales of electricity to TVA's customers in a seven-state, 80,000 square mile service area.

In 2001, 65 percent of TVA's energy production was from fossil (primarily coal) power plants; 29 percent was from nuclear, and 6 percent from hydro. Total sales were 161 billion kW hrs in 2001. The system contains over 17,000 miles of transmission lines and there is a workforce of about 13,500 personnel.

The fossil system has 11 plants, Cumberland Fossil Plant being the largest at 2600 MWe. The pilot Reliability-Centered Maintenance (RCM) programs for the TVA Fossil and Hydro Power division were conducted at the Cumberland plant.

The Cumberland Fossil Plant (CUF) is a 2-unit coal-fired steam plant capable of producing a base load of 2600 MWe (1300 MWe each unit) for the TVA grid system. The plant, which has been in operation since 1972, burns 6,000,000 to 7,000,000 tons of western Kentucky coal per year. It is located 40 miles northwest of Nashville on the Cumberland River, which serves as its raw water supply source as well as its transportation link, via barge, for both coal and limestone delivery.

As part of its commitment to compliance with the 1990 Clean Air Act, TVA constructed flue gas desulfurization (FGD) scrubbers for both units at CUF. The total project cost was more than $500 million, and operation of both scrubbers was initiated in the fourth quarter of 1994. The scrubbers are designed to remove 94 percent of the sulfur dioxide emissions in the flue gas—about 375,000 tons per year for the plant. The CUF scrubbers use the "wet process" to remove the SO_2. This process employs some 725,000 tons of limestone annually which is pulverized, mixed with water to form a slurry, and then pumped to an absorber module where the flue gas flows upwards through the slurry as the slurry flows down through the absorber vessel and recycle tank (see Figure 12.16). The chemical process, which takes place under carefully controlled density and pH parameters, converts the slurry–SO_2 mixture to high-quality synthetic gypsum (calcium sulfate) which is transferred as an effluent from the recycle tank to a holding pond. The byproduct gypsum, in turn, is saleable. Each unit has three absorber modules per unit, which are about 60 ft in diameter and 165 ft high.

The quantity and consistent high quality of the synthetic gypsum produced by the CUF scrubbers resulted in the development of a wallboard manufacturing facility

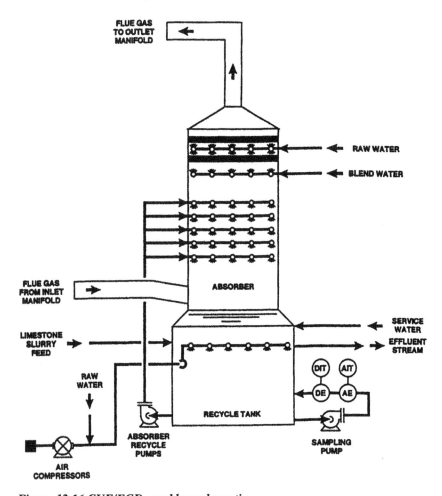

Figure 12.16 CUF/FGD scrubber schematic.

on the plant site. The manufacturing plant was built by Standard Gypsum and is one of the largest of its kind in the country, employing about 125 people. The plant, roughly 11 acres under roof, has the capacity to produce 700 million square feet of drywall per year.

Raw water is an essential ingredient in the scrubber to form both the limestone slurry and make-up or dilution water plus eliminator wash sprays in the absorber module. The scrubber uses 3.4 billion gallons of water annually, most of which is eventually returned to the Cumberland River from the gypsum effluent pond. An aerial view of the powerhouse and new scrubber facility are shown in Figure 12.17.

Figure 12.17 Aerial view of powerhouse and scrubber (courtesy of Tennessee Valley Authority).

Some Pertinent Background

Energy is a topic that has blown hot and cold periodically for at least the past 25 years. The OPEC oil crisis in the mid-1970s and the 1979 nuclear plant accident at Three Mile Island (TMI-2) were two notable events that sparked international attention and an argument about the "correct" U.S. energy policy that continues to this day. Much of this argument centers around issues of supply (e.g., dependence on foreign oil, an abundance of domestic coal), safety (e.g., nuclear accident risk), and environment (e.g., nuclear waste disposal, air pollution from fossil fuels). However, in the midst of all the rhetoric and politics involved in energy policy choices, two facts seem clear—coal is the major fuel that generates U.S. electricity, and currently available technology has greatly reduced concern about coal plant emissions (SO_2 and NO_x).

Thus, it is very timely to present this case study, which not only will discuss the efforts that TVA has taken to reduce SO_2 emissions from the largest coal plant in the TVA system, but will also describe the innovative steps taken to employ the RCM methodology to define the highly successful preventive maintenance program at the Cumberland scrubber.

In addition to successful scrubber operation at Cumberland, TVA has recently committed to spend $1.5 billion to add five scrubbers at coal plants located in Tennessee, Alabama, and Kentucky. With these additions, TVA will further reduce SO_2 emissions by at least 200,000 tons per year and bring the total sulfur dioxide emissions to nearly 20 percent below current federal standards.

System Selection and Description

When actual operating history is available, parameters such as forced outage rate, lost production output, and/or corrective maintenance costs are employed to construct a Pareto diagram for the plant systems. Such a diagram makes the "big ticket" systems very visible, and selection via the 80/20 rule becomes straight-forward. However, such data were not present in this case as the scrubber was in the final stages of construction and checkout when the RCM project was initiated. Thus, the FGD staff used their best judgment to select five possible systems for analysis, and ultimately settled on the Raw Water System because of its overriding importance in supplying one of the two ingredients that are essential for the scrubbing process.

The Raw Water System supplies both the limestone slurry preparation operation and the sprays that are an integral part of the absorber operation. In this RCM Project, the Raw Water System was further divided into six subsystems to accommodate the systems analysis process. Chief among these subsystems is Raw Water Supply, which actually brings all necessary raw water from its source to the main distribution header in the FGD facility. Thus, the Raw Water Supply Subsystem was selected for the pilot RCM project (see Figure 12.18).

The water source for the FGD is the coolant water discharge tunnels of the powerhouse which provide easy and economical access to an abundant water supply without a need to create another inlet on the river. However, these tunnels are some 20 ft below the ground level elevation of the subsystem main pumps, and this necessitated the inclusion of a vacuum priming system on the 42 in. suction piping to lift the water from the tunnels. The skid-mounted vacuum system contains two motor-driven vacuum pumps to maintain a negative pressure on the suction header. Either vacuum pump can be used, if necessary, for water pump startup or maintenance of the vacuum system. As a vacuum is developed on the system, four air release valves (one on each of three water pumps and one on the suction header) allow air in the pipes to discharge until water moves into the pipes and closes the valves. Control of the vacuum priming is automatic (via the facility distributed control system) with a local control panel for emergency and maintenance operation, and the system is alarmed to warn of vacuum loss.

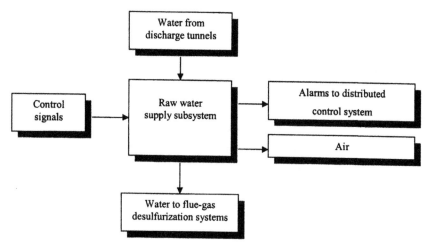

Figure 12.18 Simplified raw water supply subsystem—functional block diagram.

The three main raw water pumps discharge through automatic self-cleaning motorized strainers to two individual headers which terminate this subsystem on the outlet side of 24 in. butterfly valves where a common header is located to distribute water to the other five subsystems. The three pumps are AC motor-driven volute pumps, each with a rated capacity of 6200 gallons per minute (GPM). At full power from both units the FGD requires only 8800 GPM, so one pump is always in standby. Under certain conditions, the scrubber contains sufficient water inventory to operate for up to eight hours with total loss of raw water input. With the redundant supply mentioned above, operation could continue for up to 24 hours, thus providing a considerable grace period for corrective measures that might be required. With such an arrangement, it is highly unlikely that the scrubber water supply will ever be the cause of a forced outage at the powerhouse.

RCM Analysis and Results

The Raw Water Supply Subsystem employed the Classical RCM process described in Chapter 5. The data from the analysis process was recorded by hand onto the RCM forms (as the "RCM WorkSaver" software had not yet been developed).

This pilot project was accomplished over a seven-month time span to accommodate the scheduling of the three-person RCM team that conducted the study. The three people represented an operations shift supervisor and senior craft personnel from the mechanical and electrical/I&C maintenance group. For the pilot study, there was also an RCM consultant/facilitator and the Manager of Maintenance Programs for TVA's Fossil and Hydro Power division. The team worked one week

Number of S/S Functions	5
Number of S/S Functional Failures	12
Number of Components within S/S Boundary	46
Number of Failure Modes Analyzed	152
Number of Hidden Failure Modes	72
Number of Critical Failure Modes (i.e., A, D/A, B, D/B)	17
Number of PM Tasks Specified (including run-to-failure decisions)	156

Figure 12.19 RCM systems analysis profile.

Task Type	RCM	Conventional
Time Directed		
- Intrusive (TDI)	9	56
- Nonintrusive (TD)	1	40
Condition Directed (CD)	10	1
Failure Finding (FF)	15	1
Total PM Tasks	35	98

Figure 12.20 PM task type profile (for failure modes).

of each month for a total of 35 days of applied effort. This three-person team continued the FGD RCM program, completing three additional critical subsystems, experiencing about 25 days of applied effort per subsystem in the analysis "production" mode.

The results of the pilot project are summarized in Figures 12.19, 12.20, and 12.21. The System Analysis Profile in Figure 12.19 shows that 156 separate PM Task decisions were specified for the 152 potential failure modes identified. Almost half of these failure modes were hidden (72) but only a handful led to either safety or outage levels of criticality (17), which is a testimony to the various redundancy factors incorporated in the FGD design.

Since the scrubber was new and had no prior actual PM history for comparison purposes, the team utilized the TVA Conventional PM directives to formulate the initial scrubber PM program. Thus, this conventional PM program was used to develop the basis for comparison with the RCM results shown in Figures 12.20 and 12.21. The PM Task Type comparison is shown in Figure 12.20, and is quite revealing on two points: (1) RCM has reduced the need for PM actions by a factor

	Similarity Description	Number		Percent
I.	RCM = Conventional (Tasks are identical)	5		3%
II.	RCM = Modified Conventional (Same general task approach)	9		5%
III. A.	RCM Specifies Task Where No Conventional Task Exists (Conventional approach not cost effective)	11	√	6%
III. B.	RCM Specifies RTF Where No Conventional Task Exists (Similarity probably accidental in most cases)	74		41%
IV.	RCM Specifies RTF Where Conventional Task Does Exist (Conventional approach not cost effective)	47	√	26%
V. A.	Conventional Task Exists – No Failure Mode in RCM Analysis	20	√	11%
V. B.	Conventional Task Exists – But RCM Specifies Entirely Different Task	<u>16</u>	√	<u>8%</u>
	Total	182		100%
				√ = 51%

Figure 12.21 PM task similarity profile (for failure modes).

of 2.8; and (2) RCM has introduced a significant content of non-intrusive CD and FF tasks into the PM program. In addition, RCM resulted in 121 decisions to Run-To-Failure. The RCM Task Type Profile in Figure 12.20 also reduced the manpower requirements by a factor of three with respect to the conventional effort.

Another way to examine the RCM versus conventional PM program is to develop the Task Similarity Profile shown in Figure 12.21. This comparison reflects where the RCM-based PM tasks either concur with or differ from the conventional PM tasks. The items denoted by the check mark (√) indicate where the RCM process had its greatest impact on optimizing the PM resources by eliminating tasks that were not cost effective (Item IV) or adding tasks to cover the critical failure modes that had gone unrecognized in the Conventional program (Item III.A). Notice also that, in 20 cases, the Conventional program would conduct PM tasks where no failure mode was even identified in the FMEA (Item V.A). In 16 cases, both programs have recognized a need for PM action, but RCM took an entirely different approach to which action would be the most effective approach (Item V.B), and these differences reflect the shift from TD to CD or FF tasks. Overall, the RCM results affected a significant change to 51 percent of the original PM task plans.

RCM Implementation Process

The implementation process at Cumberland proved to be a smooth transition to the new RCM tasks. This was aided to a large extent by two factors: first, scrubber operation had not yet been initiated, so there was virtually no long-established method to overcome; and second, the maintenance superintendent conducted an all-hands meeting to explain the use of the RCM-based PM tasks and the beneficial impact RCM would have on relieving the potential for excessive OT and weekend work just to keep up with the heavy burden inherent to the Conventional PM tasks. The staff were well aware of the fact that the Conventional PM program would significantly increase their workload just to keep up with the heavy PM schedule.

The mechanics of implementing the RCM tasks were also straightforward. Cumberland uses the MPAC CMMS software. The Conventional PM program, which had been loaded, was simply purged and replaced with the RCM-based tasks. This included all of the non-intrusive CD and FF tasks.

The TVA experience with CUF/FGD would seem to indicate that implementing RCM PM tasks may be accomplished more smoothly with a new facility than with a plant that is "entrenched" in its long-standing way of doing things.

Benefits Realized

Two very important benefits were realized from the RCM program:

1. The RCM results from the Raw Water Supply Subsystem presented credible evidence that the conventional PM program that was initially used as a basis for our PM tasks was, in fact, very overstated—possibly by as much as three times! RCM provided a much more realistic estimate with which to staff accordingly.
2. The RCM results also identified the need for a redundant source of raw water to eliminate a series of plausible failure scenarios (including one potential single-point failure) that could eliminate the primary raw water source. This redundant water supply was installed to draw water from the powerhouse fly ash sluice water. To date, this backup water supply system has been used twice. The scrubber has never been the cause of a powerhouse outage.

Virtually every Classical RCM analysis produces a variety of significant (but unpredictable) benefits that go far beyond those realized from the PM task optimizations. This analysis produced a variety of "side" benefits with economic payoffs that were at least equal to the maintenance benefits. The side benefits derive from the fact that the classical RCM process is very comprehensive in its examination of the system in question, and as such compels the analysts to evaluate issues that bear on design, operation, and logistic matters of interest. A sampling of some side benefits realized in this study follows:

  a. Drainage of the vacuum tank was modified to include a "drain pot" that eliminated the loss of system vacuum when draining was carried out.
  b. Each Run-To-Failure decision was reviewed to assure that spares were available if long lead times were involved.
  c. Several areas were identified as candidates for the addition of isolation valves for on-line maintenance purposes.
  d. A "local" pressure gage was pinpointed as critical to system control, and was converted to a direct readout in the control room.
  e. The level of understanding of the functional subsystems and their interrelationships was enhanced through the RCM process.
  f. The FMEA information was used to train operators on transient and upset conditions that could occur.

Concluding Remarks

The TVA CUF PM program utilizes the Classical RCM process exclusively. The decision to use the Classical version of RCM (versus some streamlined approach) has been driven by two major considerations:

  1. RCM will be utilized only on those systems where a large ROI is possible (i.e., the 80/20 rule).
  2. The Classical RCM process historically yields significant "side" economic benefits. Such potential payoffs to the 80/20 systems must be explored to the maximum possible extent.

Use of the Classical RCM process on the 80/20 systems should be considered by anyone who may be contemplating the introduction of RCM into their maintenance strategy.

Acknowledgements

The authors wish to express their appreciation for the efforts put forth by the RCM team and supporting personnel who made this case study possible:

RCM team	*Support*
Richard Byrd	Dale Gilmore
Mark Littlejohn	Ron Haynes
Rodney Lowe	Robert Moates
Mac Smith (facilitator)	

12.2.4 Georgia-Pacific Corporation—Leaf River Pulp Operations

Brief Corporate and Plant Description

Headquartered at Atlanta, Georgia-Pacific is one of the world's leading manufacturers and distributors of tissue, pulp, paper, packaging, building products, and related chemicals. With annual sales of approximately $27 billion, the company employs more than 85,000 people at 600 locations in North America and Europe. Its familiar consumer tissue brands include Quilted Northern, Angel Soft, Brawny, Sparkle, Soft 'N Gentle, Mardi Gras, So-Dri, Green Forest, and Vanity Fair, as well as the Dixie brand of disposable cups, plates, and cutlery. Georgia-Pacific's building products distribution segment has long been among the nation's leading wholesale suppliers of building products to lumber and building materials dealers and large do-it-yourself warehouse retailers. In addition, Georgia-Pacific's Unisource Worldwide subsidiary is one of the largest distributors of packaging systems, printing and imaging papers, and maintenance supplies in North America, and is the sole national distributor of Xerox branded papers and supplies.

Located in New Augusta, Mississippi, and situated on 500 acres near the Leaf River, Georgia-Pacific's Leaf River Pulp Operations employs 345 people and annually produces 600,000 tons of Leaf River 90®, a high-quality, bleached market pulp that is free of elemental chlorine. Raw stock wood usage is approximately 3.0 million green tons per year, and the wood residue and bark supply 93% of the fuel source required for plant operation. Leaf River 90® is sold domestically and exported to Europe, Mexico, and Asia from ports in Alabama, Mississippi, and Louisiana. Goods commonly produced from the product include fine writing papers, postage stamps, tissue products, and coffee filters.

Great Northern Nekoosa (GNN) initiated plant startup in 1984, following nearly three years of construction. The construction cost of $560 million included $150 million in environmental safeguards. A state-of-the-art facility, the new mill had much to offer. Among its assets are technologically advanced equipment and environmental controls, and one of the world's largest continuous digesters. And, most importantly, there is a team of skilled, highly motivated employees fully involved in daily operational decisions. Georgia-Pacific acquired GNN and Leaf River in 1990.

Leaf River has achieved various safety awards and recognitions including the OSHA Voluntary Protection Program STAR status (its highest rating) in 1993, and the Pulp & Paper Safety Association Best Safety Record of the Year for 1998. Its environmental achievements include the production of elemental chlorine-free (ECF) pulp starting in 1990, recipient of the Mississippi Outstanding Wastewater Treatment Facility award in 2000, and National Wildlife Habitat Certifications in 1999 and 2000. The Leaf River Community Outreach Program has provided scholarships to local colleges and is the area's top contributor to United Way of Southeast Mississippi.

Seventeen years after initial startup, Leaf River Pulp Operations has exceeded its original design capacity of 1050 tons per day to produce more than 1800 tons and continues to be one of the industry's lowest-cost producers of bleached market pulp. Capital investments, including installation of a second set of evaporators and an oxygen delignification system, have played a significant role in the mill's ability to remain competitive in the global marketplace. But, by far, the greatest factor in Leaf River's continued success is a workforce dedicated to continuous improvement throughout every phase of the operation.

System Selection and Description

The Leaf River Plant is composed of two major areas of process operation that are referred to as the wet end and the dry end. The wet end takes the raw timber stock and produces a pulp slurry which is then fed to the dry end where the slurry is dried and formed into large pulp sheets. These sheets emerge from the drying process and are cut to customer requirements for packaging and shipment. The decision was made to conduct the pilot RCM project on the dry end where issues of corrective maintenance and downtime were historically more critical in meeting throughput and delivery schedules.

The dry end consists of two major process systems: the Dryer System and the Box/Pack/Ship System. Various elements or subsystems within the dry end were analyzed per Step 1—System selection of the RCM systems analysis methodology, using the most recent 12 months' data for corrective maintenance and downtime statistics to identify the 80/20 (bad actor) subsystems. This resulted in the selection of the Cutter Layboy Subsystem. A Functional Block Diagram of the Dryer System is shown in Figure 12.22.

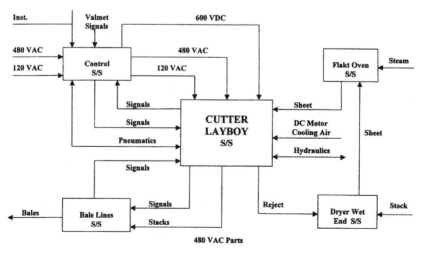

Figure 12.22 Dryer system—functional block diagram.

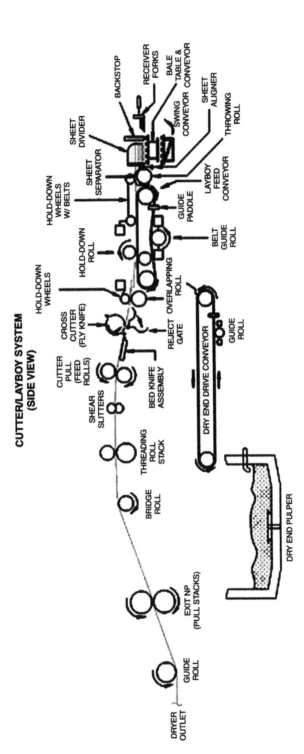

**CUTTER/LAYBOY SYSTEM
(SIDE VIEW)**

BACKSTOP

RECEIVER
FORKS

BALE
TABLE &
CONVEYOR

SHEET
ALIGNER

THROWING
ROLL

SHEET
DIVIDER

SWING
CONVEYOR

SHEET
SEPARATOR

LAYBOY
FEED
CONVEYOR

HOLD-DOWN
WHEELS
W/ BELTS

GUIDE
PADDLE

BELT
GUIDE
ROLL

HOLD-DOWN
ROLL

HOLD-DOWN
WHEELS

OVERLAPPING
ROLL

CROSS
CUTTER
(FLY KNIFE)

REJECT
GATE

CUTTER
PULL
(FEED
ROLLS)

DRY END DRIVE CONVEYOR

GUIDE
ROLL

SHEAR
SLITTERS

BED KNIFE
ASSEMBLY

THREADING
ROLL
STACK

BRIDGE
ROLL

DRY END PULPER

EXIT NP
(PULL STACKS)

GUIDE
ROLL

DRYER
OUTLET

Figure 12.23 Cutter layboy schematic.

The Cutter Layboy Subsystem is designed to cut and deliver the full width of the pulp sheet from the oven dryer to the finishing lines. The width of the web is slit into eight individual sheets. The full-width sheet is pulled through the slitters, which cut the sheet into the desired width. The sheets are then picked up by the feed roll and fed into the flyknife. (The flyknife is a rotary cutter that cuts the sheet to the desired length.) From the flyknife, the cut sheets of pulp travel to an overlapping roll. The overlapping roll, which is run at a faster speed than the flyknife, pulls the cut sheets from the flyknife and transfers them to the sheet conveyor. The speed difference between the sheet conveyor and the overlapping roll determines the amount of sheet overlap. The sheet conveyor transfers the sheets over the sheet separators, which bend the sheets. This creates a gap between the sheets, allowing the sheets to enter the boxes without hanging up on the divider plates. The sheets from the sheet separators are fed onto the throwing roll, which feeds the sheets into the boxes. The boxes consist of divider plates, sheet stops, and sheet aligners. The stacks of pulp are collected on the bale table until the predetermined count or weight is obtained, then the layboy forks collect and hold the stacks while the bale table transports them to the balelines. The Cutter Layboy Subsystem runs as long as the dryer is producing a product—usually 24 hours a day, seven days a week. The Cutter Layboy schematic is shown on Figure 12.23.

RCM Analysis and Results

The Cutter Layboy Subsystem employed the Classical RCM process described in Chapter 5. The "RCM WorkSaver" software described in Chapter 11 was used to support the effort. The RCM Project team consisted of a team leader, area operations technician, area maintenance E&I technician, mechanical maintenance technician, and the project facilitator. Also included on the team on a part-time basis were the RCM Consultant and a Corporate maintenance engineer. The RCM training was conducted during the first week of the project. All project team members, the area maintenance supervisor, and the maintenance PM supervisor attended the training. The systems analysis was conducted in weekly intervals, beginning in early May and concluding in late July 2000. This was done so that the team would not experience "burn out." Overall, the pilot project involved 25 days of team effort. The analysis was conducted in a conference room in the Dryer System area with minimum distractions. Conducting the analysis in this area allowed the team to view the selected system from time to time when questions could not be readily answered in the room. Several team members shared the duties of entering the team data onto the RCM forms via a laptop computer. The RCM software was used in conjunction with a computer Light Pro projection system for easy viewing of each entry by the whole team. The area operations and maintenance supervision reviewed and approved the final analysis information.

The results of the analysis are summarized in Figures 12.24, 12.25, and 12.26. The Systems Analysis Profile in Figure 12.24 indicates the level of detail that the team examined. Notice that nearly 9 out of every 10 potential failure modes are

- Subsystem Functions   5
- Subsystem Functional Failures   10
- Components Within Boundary   44
- Failure Modes Analyzed   198
    - # Hidden   32 (16%)
    - # Critical   172 (86%)
- # of RCM – Based Tasks Identified   249
  (including Run-To-Failure)

Figure 12.24 Systems analysis profile.

	RCM	Current
• Time Directed		
- Intrusive (TDI)	58 (23%)	29 (12%)
- Non-Intrusive (TD)	44 (18%)	7 (3%)
• Condition Directed (CD)	62 (25%)	41 (16%)
• Failure Finding (FF)	11 (4%)	0 (0%)
• Run to Failure (RTF)	74 (30%)	-
• None	-	172 (69%)
• Total Active	175 (70%)	77 (31%)
	249 (100%)	249 (100%)

Figure 12.25 PM task type profile (by failure mode).

critical (safety or outage related) in that the Cutter Layboy is a direct in-line process with very little redundancy, alternate mode, or "graceful" degradation capability. However, there is a relatively small number of hidden failures since the combination of instrumentation and operator presence makes virtually all malfunctions highly visible when they occur.

An overview comparison of the 249 proposed RCM PM Tasks with the current PM program clearly reveals the impact of this project. In Figure 12.25, we see that the active number of PM Tasks was more than doubled, from 77 to 175, with a commensurate increase in all PM Task Types (TD, CD, FF).

Figure 12.26 examines the same 249 tasks in terms of a similarity profile. That is, how much likeness or difference is there between the proposed RCM Tasks and the Current Tasks. The bottom line is that the proposed tasks change 68 percent of the current PM program. Most notably, 99 (40 percent) of the proposed RCM tasks are associated with failure modes that currently have no assigned PM task activity!

	RCM vs Current	
• RCM = Current	4	(2%)
• RCM = Modified Current	66	(26%) √
• RCM Specifies Task, Currently No Task Exists	99	(40%) √
• RCM Specifies RTF, Currently No Task Exists	74	(30%)
• RCM Specifies RTF, Currently A Task Exists	0	(0%)
• Currently A Task Exists, RCM Task is Entirely Different	6	(2%) √

$$\overline{}$$

249

Figure 12.26 PM task similarity profile (by failure mode).

RCM Implementation Process

The implementation process was structured to continue the use of the Analysis Team personnel in order to preserve a transition of first-hand knowledge about the RCM task decisions. But we also involved other mechanical and electrical maintenance technicians who had not participated directly on the Analysis Team. This provided the opportunity to familiarize them with the RCM process, and thus to gain their buy-in to the PM task changes being made. Our initial action was to obtain the approval of the mechanical and electrical lead supervisors via a thorough review of our analysis data and decisions. This approach allowed us to clarify any questions arising about the team's decisions, and in some cases to resolve the precise steps required to perform the recommended RCM-based tasks.

Our main job involved creating model work orders for the PassPort CMMS which could handle both conventional PM tasks (i.e., the TD and TDI tasks) and the more sophisticated PD or predictive-type tasks (i.e., the CD and FF tasks). Notice in Figure 12.26 that 171 of the 175 active tasks resulting from the RCM analysis are different in some respect from those of the existing PM program, and that 99 of the 171 are brand-new tasks.

It was obvious as we moved into the latter stage of implementation that we should have tackled the 171 tasks in more manageable groupings of 10–15 tasks (almost 90 percent were of Category A or B criticality) to ensure accuracy in work-order instructions, with checklists to verify correctness via a review process with the original Analysis Team. In this regard, great care must be taken with the PD task development with respect to the parameter recordings required (including consideration of the time needed for each recording), and the definition of "trigger points" that would automatically generate a PD work order from the CMMS. You should proceed to the next group of tasks only after all issues with the current group are resolved and in place.

After the PM and PD work orders were actually issued and used for a period of time, it was necessary to issue revisions due to errors in the task checklist or format, changes in task frequency, or (rarely) issue of a new task due to a failure mode event that had been missed in the RCM analysis.

Our lessons learned would include the following:

1. Assure accuracy and approval before any RCM task is actually taken to the floor.
2. Involve as many of the technicians from the floor as possible in developing the data loaded into the CMMS.
3. Keep a "core group" involved throughout the implementation process to maintain consistency in the development of work-order packages.
4. Initiate a "Living Program" to assure that changes and new information are appropriately factored into the PM and PD tasks in the CMMS.

Benefits Realized

Our implementation process constituted an initiation of a "Living Program" that identified a few failure modes that had been missed in the systems analysis, and also indicated some needed adjustments to the proposed RCM task formats.

Also, during the time period when the Cutter Layboy analysis was being conducted, we applied some of the RCM methodology and analysis details to other similar components elsewhere in the Dryer System. We then measured the results of the implementation over the last quarter of 2000 and the first quarter of 2001 and compared them to the same quarters of the previous year. On the Cutter Layboy Subsystem, the downtime was reduced by 42 percent. On the Dryer System as a whole, there was a 37 percent reduction in total maintenance dollars (since the reduction in corrective maintenance efforts far outweighed the increase in preventive maintenance efforts), and a 52 percent reduction in Dryer System downtime. On the basis of these better-than-expected benefits from our pilot project, the RCM program is being continued with both Classical and Abbreviated Classical RCM™ projects on subsystems throughout the Leaf River plant.

Acknowledgements

The authors wish to express their appreciation for the efforts put forth by the RCM team and supporting personnel who made this case study possible:

RCM team	*Support*
Jack Cross	Gene Flanders
David Hill	Charlie Hodges
Joel Wallace	Richard King
Frankie Yates	Louis Wang
Gary Ficken (facilitator)	
Mac Smith (consultant)	

12.2.5 Boeing Commercial Airplane—Frederickson Wing Responsibility Center

(*Authors' note*: This case study is based on a paper titled "RCM Comes Home to Boeing" which was presented at the Maintech South '98 Conference in Houston, Texas, and was published in *Maintenance Technology* magazine in January 2000—Ref. 6).

Introduction

Boeing Commercial Airplane (BCA) achieved a major revamping of its facility maintenance activities during the 1997–2000 period as we moved from a reactive to a proactive program. This paper will discuss some aspects of this transition and, in particular, will describe the significant role that Reliability-Centered Maintenance (RCM) has played in these efforts.

This case study—subtitled "RCM Comes Home to Boeing"—was chosen to reflect a rather fascinating story about the delayed technology transfer of the RCM process within BCA from our airplanes to the production machinery that builds and assembles these airplanes.

This story begins in the early 1960s when the Type Certification process for the 747-100 airplane was initiated by the FAA. The process required that Boeing define an acceptable preventive maintenance program for the 747-100. The FAA initially envisioned this program to be 3 times more extensive than the 707 program under the rationale that the 747 would carry 3 times more passengers. United Airlines, one of the first of two buyers (Pan Am was the other), and Boeing realized that such a requirement was so costly that the airplane could not operate in an economically viable fashion. This problem was especially amplified by the fact that the existing 707 maintenance philosophy was built on the premise that an airplane periodically "wore out," and thus required a major (and very costly) overhaul in order to retain its airworthiness stature. UA and Boeing decided to return to ground zero with a clean sheet of paper, and to challenge the validity of the wearout premise. Tom Matteson and his maintenance analysts at UA played a key role in this evaluation, and were able to use their extensive historical database to prove that, in reality, only about 10 percent of the non-structural equipment in their jet fleet showed an end-of-life or wearout characteristic.

As a result, they structured a common-sense decision process to systematically determine where, when, and what kind of preventive maintenance (PM) actions were really needed to preserve airworthiness. This new look evolved into what is known as the Maintenance Steering Group (MSG). The MSG defined an acceptable and economical PM program for the 747-100, which received FAA approval. This MSG process became the standard for commercial aviation and still exists today. In 1970, the process was labeled Reliability-Centered Maintenance or RCM by the DOD (specifically, the Navy), and is now a very popular maintenance process which is practiced in many industries worldwide.

The airplane design side of BCA, which was instrumental in the development of RCM, did not communicate this finding to the production side of the house. Rather, it took some 30-plus years for RCM to grow its roots in other industries before the Boeing facilities maintenance people learned about RCM in their Best Practices investigation. This is what brought RCM home to Boeing, and the sections to follow describe the first RCM pilot project, which took place in 1997.

Boeing Maintenance Program

Within BCA, six major regions containing all of the commercial airplane production plants have been combined under one facility organization for maintenance purposes. This organization is known as the Facilities Asset Management Organization or FAMO.

The six regions are located in Washington State, Oregon, Kansas, and California. The manufacturing centers in these regions are the main customers for FAMO's services. Region leaders report to the Vice President of FAMO, and they, in turn, have Group and Team leaders as their management structure. While there is autonomy within each region, for the most part, each of the various teams consists of mechanical, plumbing, millwright, and electrical craft personnel, maintenance analysts, and reliability engineers. Core resource groups additionally include planning, as well as equipment and plant engineering capabilities.

A key element of the FAMO management strategy was the establishment of an Asset Management Initiative in conjunction with FAMO's production customers. This initiative is a formal partnership between FAMO and production with the stated objective of significantly improving how BCA manages all of its assets in order to achieve lean manufacturing, process improvements, and dependable measures of asset utilization.

FAMO has developed a long-range strategy to optimize the application of its resources—i.e., people, material, equipment, and specialty tools. Customer (production) involvement in its execution is essential to assure that the FAMO plans align with the customer's business commitments. To achieve these goals, FAMO is deploying various state-of-the-art maintenance concepts through an Advanced Maintenance Process Program (AMAP).

BCA is investing in today's best maintenance practices, such as tactical planning and scheduling, a continuing program of craft skills development, advanced preventive and predictive maintenance technologies, TPM, and RCM. There is also an equal focus on safety and regulatory issues. One essential element of AMAP is the inclusion of the reliability parameter in all of our efforts. Within FAMO, reliability is defined as asset availability and performance, and thus it encompasses every aspect of what the organization does on a day-to-day basis to prevent the loss of critical facility systems while preserving, first and foremost, the safety of our employees.

In our review of Industry Best Practices, we learned that the RCM process used on our airplanes has been employed increasingly throughout U.S. industry as an effective decision technique to optimize the application of maintenance resources. The RCM focus has provided dramatic results in reducing equipment corrective maintenance actions and loss of system availability (i.e., less downtime). Further, when we looked at the different RCM "solutions" on the market, we also discovered that the most effective programs used the Classical RCM process which followed the original airplane methodology. This Classical RCM process was promoted almost exclusively by Mac Smith, who, literally, wrote the book on RCM (see Ref. 1). So, we elected to use the Classical RCM process, and brought Mac in as our consultant to facilitate pilot projects at three of our production sites. The first pilot project was on our Spar Mills at the Wing Responsibility Center in Frederickson (our newest factory, located south of Seattle). The remainder of our paper will discuss this Spar Mill project.

The Spar Mill RCM Project

This project was conducted at the Frederickson production facility. Two major production capabilities are located there: one is dedicated to building the vertical and horizontal composite tail sections for the 777 airplane, and the other is dedicated to producing the wing spar and skin kits that are used in all airplane types currently produced by BCA at its Renton and Everett final assembly lines. It is this latter facility where we conducted the first RCM pilot project.

The spar and skin facility (Wing Responsibility Center) is a 550-person organization with an annual operating budget of some $300 million. The FAMO personnel are composed of skilled machinists, mechanical and electrical technicians, numerical control (NC) specialists, equipment and reliability engineers, and other support personnel. This facility opened for production in April 1992. It has 21 acres of floor space that produce both spar and skin wing sections in continuous aluminum pieces up to 110 ft in length. The spar production line includes automated stringer handling, overhead crane, spar mill, drill router, deburr, paint, chip collection, shot-peen, and bending/forming systems. As discussed below, the Spar Mill System clearly met the 80/20 selection criteria for this pilot project.

The Ingersol 7-axis Spar Mill is shown in Figure 12.27. There are eleven identical spar mills on the production line; thus there is a significant benefit to be realized by the multiple application of the RCM results. Each Spar Mill is a two-spindle machine that can cut and form one large spar or two smaller opposite spars simultaneously. The mill gantry travels on ways embedded and seismically isolated in the factory floor. Each spindle has ways for horizontal and vertical

Figure 12.27 Ingersol 7-axis spar mill (courtesy of Boeing Commercial Airplane).

motion, as well as a rotational motion mode. All cutting by the mill is numerically controlled, and tolerances are typically held to 0.003 in., or less on critical cuts.

For the RCM project, the Spar Mill was divided into three subsystems: (1) Cutting, (2) Control, and (3) Auxiliary Support. The Cutting Subsystem was chosen for a detailed RCM analysis.

The RCM Team

On the basis of past success, the RCM team was composed of craft personnel (operator, mechanic, electrician), supported by a reliability engineer and a maintenance analyst. Our RCM consultant was the team trainer and facilitator. This combination of experience and hands-on knowledge of the spar mill was the key to project success since their contributions were reflective of the technical details of the equipment, and how it was used in the daily operations, and how it applied to the RCM methodology described in Chapter 5.

Initially, we had to grow as a team, since we encountered all of the long-standing norms usually found with a group of people from different disciplines who are trying to create a paradigm shift. Fear of change, suspicion of goals, ingrained viewpoints, skepticism about the "new" process, and individual agendas were all present. But, as we moved through the project, the team began to realize that we had focused on loss of function as the driving force in our thought process, and that we had been led to understand the "what and why" of equipment failure as the key to selecting a preventive maintenance action. Ultimately, the team felt a real sense of satisfaction because they were direct participants in an opportunity to provide meaningful value-added content to their daily work.

RCM Analysis and Results

The existing Spar Mill PM program was essentially a one-shot overhaul type of activity scheduled to be done on 9-month intervals. We discovered that over a five-year period since installation, this interval exceeded 9 months about 80 percent of the time, and went out to 18 months or more some 40 percent of the time. This had resulted in an excessive Trouble Call history which had then caused major downtime problems, elevating the Spar Mill to one of the 80/20 systems at Frederickson.

A statistical overview of the team's analysis and results is shown on Figures 12.28 to 12.30. These results reflect what we called the Cutting Subsystem.

The Systems Analysis Profile (Figure 12.28) provides a feel for the scope of the project which took 32 days to complete and was spread in one-week efforts over a six-month period. Notice the large number of Components involved (58) which produced 172 distinct failure modes that could deliver one or more of the 14 unwanted functional failures. Perhaps the biggest surprise here is the fact that almost half of the failure modes (42 percent) were hidden—i.e., if they occurred, the operator was unaware that any problem was developing with the Spar Mill until, at a later time, the consequence finally would show, and often in a very detrimental way. Not surprising was the fact that most (87 percent) of the failure modes were high on the criticality list, and could cause personnel injury or downtime. The team made 197 decisions on what to do with the 172 failure modes (some failure modes received multiple PM actions).

Turning now to Figure 12.29, we see the makeup of the RCM results in the task type profile, and its comparison to the existing "CPM" as it was called. The outstanding point here is that the number of failure modes receiving no attention currently (120) was reduced by 1/3, and the number of non-intrusive actions with condition-directed (including PdM) and failure-finding tasks was increased fourfold.

But the real impact of the RCM results is seen in Figure 12.30, the task similarity profile, where we examine the similarities and differences between the RCM

Spar Mill #6 – Cutting Subsystem

- Subsystem Functions   8
- Subsystem Functional Failures   14
- Components Within Boundary   58
- Failure Modes Analyzed   172
  - # Hidden   72 (42%)
  - # Critical   150 (87%)
    (i.e. A, D/A, B, D/B)
- # of RCM – Based Tasks Identified   197

  (including Run-To-Failure)

Figure 12.28 Systems analysis profile for spar mill #6—cutting subsystem.

Spar Mill #6 – Cutting Subsystem

	RCM	Current
Time Directed		
- Intrusive (TDI)	32 (16%)	34 (17%)
- Non-Intrusive (TD)	16 (8%)	30 (15%)
Condition Directed (CD)	34 (17%)	2 (1%)
Failure Finding (FF)	21 (11%)	11 (6%)
Run to Failure (RTF)	84 (43%)	-
None	-	120 (61%)
Design Modification (D)	10 (5%)	0 (0%)
	197	197**

Figure 12.29 PM task type profile (by failure mode) for spar mill #6—cutting subsystem.

and Current PM tasks. Since we knew from our analysis just where the critical failure modes were located, we were able to develop appropriate PM actions where needed (see Item 3A in Figure 12.30), and delete PM work where not needed (see Items 4 and 5A in Figure 12.30). Overall, the RCM results changed 71 percent of the current "CPM" on the Spar Mill. Since there are 11 such Spar Mills at Frederickson, the multiplying effect here is quite dramatic.

Finally, the intensity of the RCM process enabled the team to discover a number of non-maintenance-related findings, which also provide a significant portion of

Spar Mill #6 – Cutting Subsystem

		RCM vs Current	
1.	RCM = Current	0	(0%)
2.	RCM = Modified Current	45	(21%) √
3A	RCM Specifies Task, Currently No Task Exists	56	(26%) √
3B	RCM Specifies RTF, Currently No Task Exists	64	(29%)
4.	RCM Specifies RTF, Currently A Task Exists	18	(8%) √
5A	Currently A Task Exists, but no failure mode was identified	22	(10%) √
5B	Currently A Task exists but the selected RCM Task is Entirely Different	14	(6%) √
		219	

Figure 12.30 PM task similarity profile for spar mill #6—cutting subsystem.

the benefit achieved. We call these findings IOIs or Items Of Interest. There were 36 IOIs recorded, and they affected design, operations, reliability, safety, and logistics.

Implementing the Analysis Results

Implementation of the PM Task findings that were developed in Step 7 of the System Analysis Process proved to be challenging. There were several reasons for this:

1. We needed to develop a general understanding of the RCM process and a buy-in to the results of the analysis from a broad group of personnel who were peers of the team members. Nothing new on a factory floor is ever successfully deployed by simply announcing that it will happen!
2. Several of the new tasks required a more direct participation on the part of the operators, and we needed to carefully integrate this with the production shift supervisors.
3. With the substantive changes being made to the current "CPM" procedure (71 percent), we had to take the time to develop several new procedures and coordinate their review and approval with all affected parties.
4. Several of the condition-directed tasks required some exploratory work to ascertain their suitability for the failure mode(s) in question. Some of this work is still ongoing.

All of the above required extensive communication across organizational lines and among the many disciplines that are resident in the production and maintenance workforce. You will notice that involvement of production personnel, together with a close integration of their viewpoints and experience with FAMO, was a key ingredient in the entire project.

At this time (early 2000), we have deployed 21 RCM-based PM procedures on the Spar Mill. These procedures essentially encompass all of the analysis findings, except for a few condition-directed tasks still under evaluation. The new PM format being used includes additional descriptions of the work to be performed plus references to the specific failure modes and failure causes that triggered the PM tasks. Deployment to the factory floor was done in a stepwise fashion in order to introduce the shift from traditional to RCM tasks without disruption. Again, open communication was essential to successful implementation. Giving honest and positive feedback to the many questions that were asked was crucial to creating a positive paradigm shift.

We are also in the process of evaluating and, where appropriate, implementing the IOIs. To date, several have been accomplished, including, as examples, the following:

- Pressure washing of the entire machine is being eliminated in favor of selective washing of a few components. This will virtually eliminate severe corrosion and chip contamination damage caused by the pressure wash.
- Spindle vibration analysis is being closely correlated with the as-produced part quality (tolerance) to obtain the maximum spindle life before change-out.
- A & B axis rack covers have been removed since they trap chips and cause pinion seal damage, rather than prevent chips entering the racks.
- In the future, all seven axis drive motors will be replaced with brushless motors, thus eliminating five specific failure modes of concern.
- Counterbalances are being added to all W & Z axes to eliminate failure of the thrust bearing.

Return-On-Investment (ROI) Considerations

Our objective with the RCM program is to focus our PM resources in order to reduce costly corrective maintenance actions and resulting loss of machine uptime. While no hard measurements are yet available, we can make the following observations at this time:

1. While the PM program has changed significantly, its cost is virtually unchanged. Costly TD tasks have been replaced by

inexpensive CD and FF tasks, and the task frequencies have been extended.

2. With the program now focused on the critical failure modes, we expect to see a reduction in unexpected corrective maintenance actions (i.e., trouble calls) of at least 50 percent. Downtime should also decrease by at least 50 percent.

3. From preliminary analyses, we know that our IOIs, when implemented, have the ability to produce annual savings in excess of $3 million. Of those incorporated to date, annual savings in excess of $0.5 million are expected.

Future Directions

With the experience gained and success achieved at Frederickson, we have already initiated and recently completed two additional RCM pilot projects at Everett and Wichita. These projects were performed on a Cincinnati 5-axis router and Modig extrusion mill respectively, and are currently in the transition to implementation. We expect to see ROI benefits similar to those at the Spar Mill accrue from these projects.

Currently, we plan to initiate additional projects at Wichita on critical systems for the production line where BCA builds all 737 fuselages as well as the cockpit portion of the fuselage for all other Boeing airplanes.

We also intend to invoke the RCM Living Program on all completed projects in order to periodically update the PM tasks as may be required, and to effectively measure the results of the RCM program.

We intend to continue the use of the Classical RCM process on our critical systems because the actual benefits have exceeded our original expectations.

Acknowledgements

The authors wish to express their appreciation for the efforts put forth by the RCM team and supporting personnel who made this case study possible:

RCM team	*Support*
Dave Bowers	John Donahue
Bob Ladner	Dean Nelson
Pat Shafer	Max Rogers
Lynn Weaver	Dennis Westbrook
Mac Smith (facilitator)	

12.2.6 Arnold Engineering Development Center (AEDC)—Von Karman Gas Dynamics Facility

Corporate and Facility Description

AEDC is a national resource with an estimated capital investment of over $6 billion. It is a very large facility with a complex infrastructure of equipment and systems, such as compressors, exhausters, cooling and refrigeration systems, air heating systems, and the most advanced computers and data acquisition systems. Some of the compressors are capable of providing very large amounts of air at pressures up to 4000 psi. Some 56 large electric motors, delivering more than 1.3 billion horsepower, drive these large and complex facilities. Several of these systems are old, dating from the 1940s, and are still being effectively used and maintained today. Figures 12.31 and 12.32 outline the Center's capabilities and complex infrastructure, and Figure 12.33 models a typical test facility or plant.

To get a true sense of the immensity and complexity of the Arnold Engineering Development Center, one needs to understand that no single facility in the world has in one place the same range of sophisticated technologies and flight simulation test facilities as those encompassed at AEDC. Within its boundaries are 58 separate

Aeropropulsion Systems Test Facility (ASTF)	Engine Test Facility (ETF)	***Von Karman Gas Dynamics Facility*** (VKF)	Propulsion Wind Tunnel Facility (PWT)
90,000 lbs Freon Refrigerated Systems	1940's Vintage Compressors (B-Plant)	**10 Compressors in series – 4000 psi**	30 Foot Diameter Compressor
1500 lbs-m/sec Compression	60,000 lbs Freon Refrigeration	**Electric Motors Total 92,500 Hp**	Electric Motors up to 83,000 Hp each
2200 lbs-m/sec Exhaust Capacity	Liquid Air System		
Electric Motors Totaling 215,000 Hp	Multi-Stage Cooling		
387,000 gal H₂O/min Cooling	Fuel Temperature Conditioning		
1700 Steady State & Transient Parameters			
1 Billion BTU/hr Heater Capacity			

Figure 12.31 AEDC facilities.

MAJOR SYSTEMS	A PLANT	B PLANT	C PLANT	P PLANT	V PLANT	TOTAL
Main Drive Power (Hp)	138,000	48,000	609,000	398,000	**100,000**	1,293,000
Maximum Motor Size (HP)	36,000	6,000	52,500	83,000	**16,000**	83,000
Number of Motors ->3,000 Hp	8	12	18	11	**7**	56
Maximum Airflow (lbs/sec)	500	200	2,200	15,500	**84**	N/A
Maximum Pressure (psia)	120	50	150	30	**3,800**	N/A
Refrigeration (Tons)	4,900	3,060	23,000	1,500	**600**	33,060
Number of Main Compressors	5	10	18	18	**15**	66
Main Ducting - Wind Tunnels (Miles)	2	2	3	11	**10**	28
Hydraulic Systems (psi)	10	8	18	21	**10**	67

Figure 12.32 AEDC major systems.

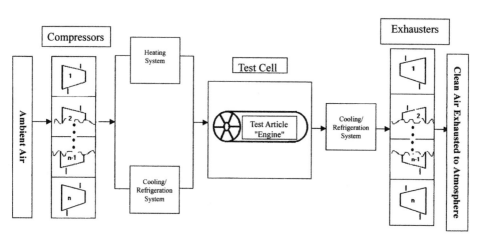

Figure 12.33 Typical test plant at AEDC.

and, in many cases, one-of-a-kind test facilities composed of aerodynamic and propulsion wind tunnels, rocket and turbine engine test cells, space environmental chambers, arc heaters, ballistic ranges, and other specialized units. These facilities can simulate flight conditions from sea level up to an altitude of 300 miles (near space), and velocities ranging from subsonic to Mach 20 (20 times the speed of sound).

The mission of AEDC is to test and evaluate aircraft, missile, and space systems and subsystems at the flight conditions they will experience during their operational life. In meeting its mission, AEDC's role is to facilitate customers in the development and qualification of flight systems, to prove and improve system designs, to establish pre-production performance benchmarks, and to assist its customers with problem recognition, analysis, resolution, and correction of current operational systems.

AEDC is located in middle Tennessee at a former U.S. Army Air Corps base now under the direct authority of the U.S. Air Force. AEDC is so named in honor of General Hap Arnold whose foresight saw the need to establish a centralized site where German technology captured during the Second World War could be evaluated, and new technology could be developed and tested that would take the United States into the space age. AEDC has evolved from its government top-secret mission into one that benefits both the armed forces and industry alike, where companies like General Electric have tested engine designs used to power the Boeing 777 airliner. To accomplish this expanded mission, the U.S. Air Force employs civilian contractors, one being Sverdrup Technology, Inc., to maintain and operate AEDC's unique facilities. Today AEDC finds it will no longer be completely funded by Department of Defense budgets, but must increasingly sustain itself as a commercial and industrial test facility.

It is the need for this equipment to be cost competitive in an increasingly expanding commercial venture that is one of the primary drivers of the current RCM program at AEDC under the control of Sverdrup Technology. AEDC no longer has the luxury of operating solely on government funds, but must now also win competitive commercial contracts that require it to be reliable, available, and profitable. AEDC is progressively realizing this goal through the application of RCM to reduce the overall cost of maintenance and unanticipated downtime, thereby increasing profitability.

Some Pertinent Background

RCM at Arnold Engineering Development Center (AEDC) has for many years been a work-in-progress. As with most companies, the thought of applying RCM has waxed and waned at AEDC, and those responsible for operations and maintenance have struggled trying to improve the bottom line, i.e., reduce the operating costs and increase profitability. Several attempts at implementing an RCM program in one form or another were made over the years dating from 1985, with an effort to decrease the number of PM actions.

The 1985 effort assumed that PM was not very effective since the total cost of maintenance remained high, i.e., a significant number of equipment failures still occurred even though a PM program was in place. It also assumed, incorrectly, that the problem, at least the cost portion of it, would be corrected by reducing the number of PMs. PM reduction took place with very little regard for why *(the need to preserve function)* and initially appeared to save money ($$$). Saving money was, in fact, the primary driver of this program, and ultimately proved to be the effort's undoing, since total maintenance costs continued to rise. This initial attempt at optimizing the PM program was primarily a system engineering approach that isolated those making the decisions, the maintenance engineers, from those who experienced first hand the results of a poor PM program, the maintenance craft technicians.

A second attempt at RCM was made in 1992 that did employ some of the basic concepts fostered in this book by the authors. That approach too had a fatal flaw, the flaw being that this again was an engineering approach and not a total team effort taking advantage of what the craft technicians knew, so ultimately the analysis was deemed questionable and implementation floundered.

Then, starting in fiscal year 1997 (FY97), the U.S. Air Force committed major resources to improve its maintenance processes at AEDC. Since then the AF/DOO office and Sverdrup Technology have been mutually charged with decreasing the cost of maintenance and increasing the availability of test units and support equipment. This is where we pick up the details of the AEDC Case Study.

In FY98 several different RCM processes were evaluated to determine the "best of the best." From this effort many lessons were learned regarding the applicability of the different RCM processes along with the acceptance of the process by the equipment and systems experts—the operators and maintainers. In FY99 and beyond Sverdrup Technology focused on one RCM process, the one that was and is the process presented in this very book by the authors (Chapters 5 and 6).

System Selection and Description

In the FY97/FY98 time frame, the decision was made to initially undertake several RCM pilot projects in parallel in both the Propulsion Wind Tunnel and Von Karman Gas Dynamics Facilities. Three different versions of the RCM analysis process were employed in these pilot projects, one being the Classical RCM process described here in Chapter 5, which was used on one system each from these two facilities.

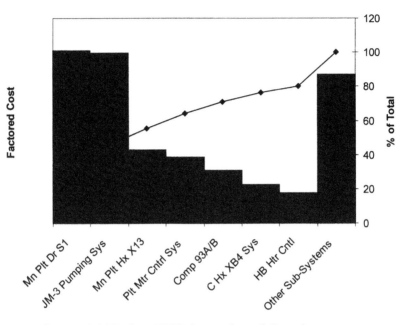

V (VKF) Plant Factored Cost Failure Data FY96-FY97

Figure 12.34 V plant (VKF) factored cost failure data.

For purposes of the Case Study, the pilot project from the Von Karman Gas Dynamics Facility (i.e., the V plant) was chosen for this discussion. As noted above, the Classical RCM process was ultimately selected, after a comprehensive review process, to be the process of choice for the program in FY99 and beyond.

Cost data for corrective or unscheduled maintenance (CM) for the Von Karman Gas Dynamics Facility, denoted as VKF or the V plant (refer to the highlighted areas of Figures 12.31 and 12.32 for additional information) for the period of FY96 through FY97 were supplied by the Sverdrup Technology quality office. These data were then arranged in a Pareto diagram and the 80/20 rule was applied to discover the *bad actor* systems (a representation of this data is found in Figure 12.34). The JM-3 Pumping System was selected as the candidate RCM system. However, the JM-3 System is very complex and wide ranging, with an array of compressors taking ambient air and sequentially compressing it until a pressure of 3800 psi is reached. This complexity required a further breakup of the JM-3 System into smaller, more manageable portions. AEDC plant failure data indicated that the <u>C92 compressor</u> was a major contributor to CM actions in the JM-3 Pumping System, so it was selected as the target RCM system.

The C92 Compressor is a Subsystem of the JM-3 Pumping System and comprises a single Ingersoll-Rand CENTAC centrifugal air compressor driven by a General Electric 1250 Hp, 6.9 kV, 3600 rpm motor. The rotating shafts and bearings of both the compressor and motor are lubricated from an oil reservoir underneath the common mounting skid. The compressor takes in filtered atmospheric air and compresses it through three stages, developing 6050 cfm at 100 psig. The 100 psig air is then reduced by passing it through a throttling pressure control valve to deliver 38 psig air to the next set of compressors. (Note: The C92 Compressor is one of a group of compressors of the JM-3 Pumping System sequentially connected to deliver a final air pressure of 3800 psig. The compressed air is then used across the AEDC facility in the various wind tunnel test cells, creating the test condition described earlier in this Case Study.)

RCM Analysis Process and Results

As noted previously, some very hard but important lessons were learned in the early attempts to develop an RCM program. These lessons included the necessity to form RCM Analysis Teams that included not only the responsible System Engineer, but also knowledgeable craft personnel representing both operations and mechanical, electrical, and I&C maintenance. This combination proved to be the key to success. (The authors served as Team Facilitators on several of the projects.)

The early pilot studies included a 3-day training session for team members, and a series of 1-day seminars for AEDC staff personnel to acquaint them with RCM and the future changes that would likely develop. These 1-day seminars eventually touched virtually every manager, engineer, and technician at AEDC.

The pilot projects involved about 25–30 days of team effort, and were staggered in 5-day increments over about a 3–5 month interval to avoid any major disruption to the normal O&M schedules. A learning curve also quickly reduced the analysis time by 20% or more, and later systems employing the Abbreviated Classical RCM™ process (see Sec. 7.2) reduced the analysis time by 50 percent. All projects conducted after the first round of pilot projects employed the RCM WorkSaver software (see Chapter 11).

The results of the analysis are summarized in Figures 12.35, 12.36, and 12.37. The System Analysis Profile in Figure 12.35 indicates the extent of effort and detail that was realized in that 231 separate failure modes in 60 different components were identified and evaluated. About one-third of these failure modes were "hidden" and all but a small fraction were labeled as "critical" (84 percent). This led to 254 distinct decisions to specify a PM task, including RTF decisions and in some cases multiple PM task actions. The specific mix of the 254 decisions is shown in Figure 12.36 by Task Type. The major points of note here are the significant reduction of failure modes in which nothing was being done (150 to 94)

- Subsystem Functions   6
- Subsystem Functional Failures   12
- Components Within Boundary   60
- Failure Modes Analyzed   231
    - - # Hidden   75 (32%)
    - - # Critical   193 (84%)
- # of RCM – Based Tasks Identified   254
  (including Run-To-Failure)
- # of Items of Interest   31

Figure 12.35 C92 systems analysis profile.

	RCM	**Current**
• Time Directed		
- Intrusive (TDI)	53 (21%)	43 (16%)
- Non-Intrusive (TD)	50 (20%)	47 (18%)
• Condition Directed (CD)	39 (15%)	17 (6%)
• Failure Finding (FF)	16 (6%)	9 (3%)
• Run to Failure (RTF)	94 (37%)	- - -
• None	- - -	150 (57%)
• Design Modification (D)	2 (1%)	0 (0%)
	254*	266*

Figure 12.36 C92 PM task type profile (by failure mode).

		RCM vs Actual	
1.	RCM = Current	4	(2%)
2.	RCM = Modified Current	72	(28%) √
3A	RCM Specifies Task, Currently No Task Exists	59	(23%) √
3B	RCM Specifies RTF, Currently No Task Exists	88	(35%)
4.	RCM Specifies RTF, Currently A Task Exists	6	(2%) √
5.	Both Specify a Task, But RCM is Entirely Different	25	(10%) √
		254	

Figure 12.37 C92 PM task similarity profile (by failure mode).

and the dramatic increase in non-intrusive CD and FF PM tasks from 9 percent to 21 percent of the total. In addition, it was found that 12 tasks in the Current PM program (266 minus 254) involved tasks that could not be identified with any failure mode or were failure modes receiving the same PM action from two different organizations.

But perhaps the most striking result is shown in Figure 12.37, the PM Task Similarity Profile. Of particular note is the fact that 59 failure modes resulted in the specification of a PM task where none existed in the current program. Overall, the RCM analysis changed the content of the Current PM plan by 63 percent.

RCM Implementation Process

At AEDC it was recognized early on that implementing the RCM recommendations, not only from this one study but from the 50 other studies completed through FY2001, required the development of a formal process to implement a change in how "things" got done. No one person, let alone a single department, could possibly be expected to complete such a monumental task—it was to become a cross-organization objective and responsibility if the entire effort was to be successful.

It should be noted that both the Classical RCM and the Abbreviated Classical RCM™ approaches produce tasks at the failure mode level which are then combined or rolled-up into higher level PMs that become the facility's preventive maintenance program. In this instance, the 160 failure mode-level tasks were combined into 15 facility-level PMs, i.e., CMMS scheduled events.

The basic implementation process, shown in Figure 12.38, requires the engineer primarily responsible for the analyzed system to categorize implementation needs into short-, medium-, and long-range items. Typical short-range items would be the modification of an existing PM task where only a few items, such as the task interval, were to be changed. Medium-range items comprised the bulk of the new work where new PMs and their corresponding work instructions had to be written. The long-range items were mostly the IOI recommendations that required additional engineering evaluation before they could be considered or implemented. Some mid-range items that had long lead-times before they would be needed, e.g., overhauls, were also placed in the long-range category.

To support the implementation process, a monthly status reporting scheme was developed. As each of the items was completed, it was immediately put into use and has become part of the maintenance program documentation.

The greatest area of difficulty, if indeed it could be so classified, was a tendency to revert back to the old way of thinking. RCM was seen as a journey. Setbacks, such as early or unanticipated failures, would occur. However, these setbacks were

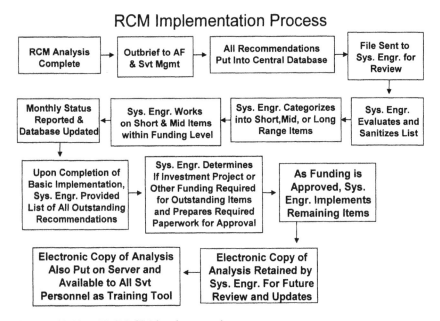

Figure 12.38 AEDC RCM implementation process.

seen as an opportunity to improve the strength of the analysis, and to begin the RCM Living Program. They were not allowed to taint the RCM program's success.

As with any program, there were lessons learned. Perhaps the hardest was that "RCM is a journey" and cannot be rushed. Though AEDC lives in a world that demands fast turn-arounds and compliance to satisfy themselves and especially their customers, AEDC and Sverdrup Technology realize that a cost-effective and applicable preventive maintenance program is worth the effort and the time required to achieve it.

Benefits Realized

The C92 Compressor RCM study was completed near the end of the 1998 fiscal year and, as with all PM improvement programs, there was a lag between completion of the analysis and full implementation of the RCM study recommendations which were completed in mid to late FY99. Since time—and especially trouble—waits for no one including RCM, the C92 Compressor experienced a major failure in FY99 that, while recognized by the RCM analysis, had not been fully implemented at the time of the problem. The occurrence of this problem and the subsequent improvements to the PM program are shown in the reduction of

C92 Compressor
PM / CM Factored Costs

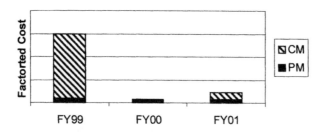

Figure 12.39 C92 compressor—PM/CM factored costs.

the corrective maintenance (CM) events displayed in Figure 12.39 for fiscal years 2000 and 2001.

As was pointed out earlier, Items Of Interest (IOIs) are *pearls of added value,* or improvement opportunities, which are uncovered anytime throughout the RCM analysis but are not directly a part of the formal task analysis process. The analysis of the C92 Compressor produced 31 IOIs. Twenty-nine IOIs were suggestions for design change or modification to the physical plant that were viewed to improve reliability, maintainability, or operability. Two IOIs suggested improvements in the stores (spare part inventory) or logistical areas (getting spare parts to the field where and when they were needed). While the calculation of a reasonable Return-On-Investment (ROI) for these IOIs is beyond the scope of this case study, it can be stated without any reservation that, when combined, these 31 *Pearls of Added Value* have the potential to generate yearly cost savings in excess of several million dollars.

Acknowledgements

The authors wish to express their appreciation for the efforts put forth by the RCM team and supporting personnel who made this case study possible:

RCM team	*Support*
Ed Ivey	Paul McCarty
Brown Limbaugh	Bert Coffman
Brian Shields	Dan Flanigan
Ronnie Skipworth	Ramesh Gulati
Glenn Hinchcliffe (facilitator)	

12.2.7 NASA—Ames Research Center/12 ft Pressure Tunnel (PWT)

(Author's note: This Case Study was the pilot RCM project at Ames Research Center, and was conducted in the 1995–1996 period. Subsequently, two additional systems were completed on the 12 ft PWT, two systems were completed on the 40 ft × 80 ft × 120 ft Wind Tunnel, and five systems were completed on the Unitary Plan Wind Tunnel. The Wind Tunnel RCM Program was completed in 1998.)

Some Pertinent Background

At the mention of NASA in the public media, the popular perceptions immediately turn to the Space Shuttle, Space Station, unmanned weather and communications satellites, research aircraft and the like. Lesser known, however, but in reality at the core of NASA's efforts and successes, is a series of Centers throughout the U.S. that conduct the science and engineering efforts necessary to produce these more visible results. These Centers (such as Ames, Langley, Lewis, Kennedy, Johnson, Marshall, and others) are composed of a wide variety of complex research and development facilities and equipment which require extensive care and maintenance to assure their safe and productive functioning.

NASA Headquarters has been promoting an agency-wide effort aimed at reducing the operating and maintenance costs at NASA facilities by sharing resources, experiences, and practices among the Centers. Central to this effort has been the decision to employ the Reliability-Centered Maintenance (RCM) methodology as a focus for achieving increased efficiencies and reduced costs in the NASA maintenance programs. The Ames Research Center is playing a key role in this effort, and this paper discusses the initial pilot project at Ames in this RCM program.

Center and Tunnel Description

NASA-ARC, located in Mountain View, California, was established in 1939 and is known internationally for its capabilities and achievements in the field of developmental aeronautics—both computational and experimental. The Center has a variety of ground test facilities which include several wind tunnels and one of the largest supercomputer facilities in the world. Its facilities include the National Full Scale Aerodynamics Complex—a 40 × 80/80 × 120 ft atmospheric wind tunnel (the largest wind tunnel in the world), and the 12 ft pressure wind tunnel (PWT), which is the subject of this Case Study.

The 12 ft PWT returned to operation in 1996 after a multi-year $120 million reconstruction effort. It is capable of operating at pressures that range from 0.14 to 6 atmospheres, thus providing a capability to test a single model configuration over a wide range of Reynolds numbers. The "new" 12 ft tunnel contains a special feature which permits pressure isolation of the plenum section (which contains the test section) from the tunnel circuit. This feature enables model changes to be made by reducing to atmospheric pressure in only the plenum section while the rest of the tunnel circuit remains at the selected operating pressure. With this feature, access time from 6 atmospheres is 20–30 minutes rather than the 2-hour interval previously experienced when the entire tunnel was depressurized and repressurized.

System Selection and Description

Since this was a new facility with no PM or CM history, we elected to construct a Pareto diagram based on information defining planned hours for the conventional PM tasks. The tunnel is composed of 12 unique systems, and the available PM task planning data for the Makeup Air System (MU) was used as a baseline. All other systems were assigned estimated annual PM hours as a function of their complexity relative to MU. This approach resulted in the information displayed in Figure 12.40. The Plenum System (PL), which ranked first, was selected for use in the RCM pilot project. This selection had the added benefit of subjecting this new and unique design to the detailed analysis that occurs in the RCM Systems Analysis Process.

The Plenum System design, which provides this quick test model change capability, is comprised of a carousel (or turntable) that rotates the test section 90° between the run position and the access position, and 15 ft isolation doors (valves) to seal the tunnel circuit upstream and downstream of the test section. A similar access door (valve) can be opened after blowdown for entry to the test

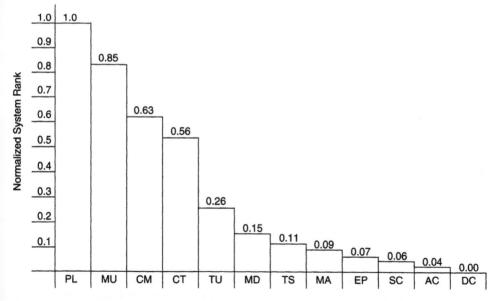

Figure 12.40 12 ft pressure wind tunnel—normalized ranking by estimated annual PM man hours.

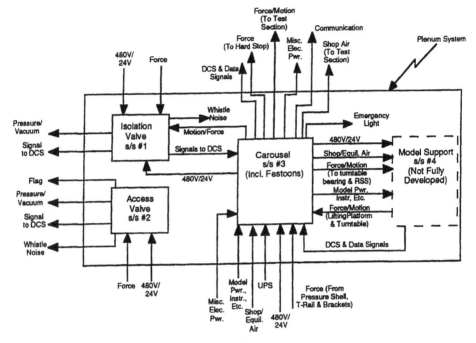

Figure 12.41 Functional block diagram of plenum system.

section. The design involves a complex arrangement of structures, motors, gearboxes, and drive mechanisms that rotate the carousel and move the three doors in and out of position, and includes locking and sealing devices on the doors. A variety of instrumentation controls these movements, assures proper positioning, and protects the facility from accidental pressure release. In order to maintain the Plenum System in top-notch working condition, thus avoiding loss of production test time and preserving all safety features, it was necessary to install a well-designed preventive maintenance (PM) program for all critical tunnel systems, the Plenum System being at the top of that list.

Due to the complexity of the Plenum System, we divided it into four Functional Subsystems. This is shown in the Functional Block Diagram in Figure 12.41. The RCM pilot project was scoped to include only the Access Valve, Isolation Valves, and Carousel Subsystems. This paper reports on the Carousel Subsystem analysis and results.

RCM Analysis and Results

The 12 ft PWT pilot project was conducted with two objectives in mind: (1) to demonstrate the benefits available to ARC from the Classical RCM process; and

(2) to train a team of wind tunnel operations and maintenance personnel on how to perform the 7-step system analysis process. Both objectives were successfully achieved. However, due to turnover in personnel who were on the original team, additional training for members subsequently assigned to the follow-on teams was required.

AMS Associates (Mac Smith) was retained as a Consultant to guide and facilitate the pilot project. The RCM team for the project was composed of two highly experienced craft supervisors, a maintenance engineer, and a facility engineer. As the Consultant noted, use of craft personnel on an RCM team has produced some of the most successful RCM programs because of both extensive knowledge of the facility that they bring to the table and an ownership of the process by the very people who are now expected to execute the results.

The pilot project took 39 days to complete, including the development of specific maintenance task recommendations. These 39 days were staggered in one-week increments over a seven-month period in order to preclude any major conflicts or discontinuities in the team's regular work assignments. In the future, the schedules for RCM projects may be compressed somewhat closer, but, in general, we find that some form of staggered intervals works well for a team which includes craft personnel. With the training process now successfully achieved, we expect that future projects will be done, on average, in 20 days, as was demonstrated on the subsequent 40 × 80 × 120 ft and Unitary Wind Tunnel studies.

The analysis documentation for the pilot project was done manually. Some of the later projects utilized the RCM WorkSaver software which became available.

In Figure 12.42, the PM Task Type Profile compares the Conventional PM tasks scheduled in the CMMS with the RCM tasks developed in Step 7 of the analysis. The striking feature of this comparison is that, while the total number of PM tasks

Task Type	RCM	Current
• Time Directed		
- Intrusive (TDI)	11	23
- Non-Intrusive (TD)	14	18
• Condition Directed (CD)	15	0
• Failure Finding (FF)	12	5
• Run to Failure (RTF)	18	N/A
• None Specified	N/A	39
	70	85

Figure 12.42 PM task type profile.

		RCM		%
1.	RCM = Current (Tasks are Identical)	4		5%
2.	RCM = Modified Current	16	√	19%
3A	RCM Specifies Task, Currently No Task Exists (Current Missed Important Failure Modes)	24	√	28%
3B	RCM Specifies RTF, Currently No Task Exists (Similarity Probably Accidental in Most Cases)	15		18%
4.	RCM Specifies RTF, Currently A Task Exists (Current Approach Not Cost Effective)	3	√	3%
5A	Currently Task Exists - No Failure Mode in RCM Analysis	9	√	11%
5B	Currently Task Exists – RCM Specifies Entirely Different Task	14	√	16%
		85		100%
			√	77%

Figure 12.43 PM task similarity profile.

is quite similar (46 versus 52), the mix of task type is dramatically altered. Specifically, the number of intrusive time-directed tasks has been reduced by a factor of two while the number of non-intrusive condition-directed and failure-finding tasks was increased by a factor of five. This is projected to reduce the number of human errors in the PM program that so frequently occur with intrusive actions, while effectively identifying required maintenance actions before they result in costly surprises and outages.

Another way to examine the RCM versus current PM structure is shown in Figure 12.43, the PM Task Similarity Profile. This comparison reflects where the RCM-based PM tasks either agree with or differ from the Conventional PM tasks. The items denoted by the check mark (√) indicate where the RCM process had its greatest impact on optimizing PM resources. Overall, the RCM process introduced a change to 77 percent of the Conventional PM program for the Carousel Subsystem. The majority of this change occurs in Items 3A (28 percent) and 5B (16 percent), where RCM either introduced PM tasks where nothing existed or opted for a more effective way of doing the PM activity.

RCM Implementation

Prior to initiation of the Carousel Subsystem RCM pilot project, work had progressed almost to completion on the loading of the conventional PM tasks into the

Maximo CMMS. Thus, the key to achieving a successful implementation of the RCM-based PM Tasks was the deletion of these Conventional PM tasks (except for the four tasks where no change was required), and the loading of the new RCM tasks into the CMMS. As others have learned, until this adjustment to the CMMS is achieved, the RCM analysis results are destined to languish on the shelf and perhaps be lost in antiquity.

In order to make the adjustment to the ARC CMMS, several actions were required to translate the brief PM task descriptions in Step 7 of the system analysis results into the complete set of data required to generate work orders and procedures from the CMMS. Absent this information, on-the-floor implementation was not possible within the standard modus operandi at ARC. These actions included the following:

1. Each task entry into the CMMS required the generation of some 20 items of data (e.g., equipment ID and description, job plan, tool/material requirements, craft type and man-hours, special requirements such as lockout/tagout, procedures needed, etc.).
2. The Carousel Subsystem boundary and equipment list in the RCM analysis differed somewhat from the conventional definition, and needed adjustment to avoid gaps or overlaps in other systems in the CMMS.
3. Every RCM work order needed a system specialist, electrical supervisor, and/or mechanical supervisor to define the correct data. The availability of these personnel was limited and required some careful scheduling to obtain their inputs.
4. Frequently, the RCM task required a completely new or modified written procedure. The system specialist and RCM team members coordinated their inputs to assure that the task intent was fulfilled properly.
5. In dealing with the condition-directed and failure-finding tasks, most of which were new (see Figure 12.42), craft training and new equipment acquisition were necessary.
6. All RCM work orders received a review and approval from the RCM team before final management approval.

We essentially found that the implementation process outlined above worked quite well, but was more involved than originally anticipated. This was primarily the result of the significant changes that were recommended by the RCM analysis process versus our original expectation that the changes would be rather minimal. We have since learned from others, as well as from our own follow-on projects, that changes in the 50 to 75 percent area are quite common. Clearly, the message that we see conveyed by this latter statistic is a significant indicator of the need to improve our conventional or traditional ways of defining equipment PM tasks. Also, it warns us not to underestimate the significance of effort required for the implementation of new PM tasks (be they RCM-based or otherwise).

Benefits Realized

In terms of benefits achieved from an RCM project, the savings in PM man-hours, overall, were small compared to the expected major economic benefits that accrue in these three areas:

1. Corrective maintenance (CM) costs were dramatically reduced owing to the effectiveness that resulted from the RCM process. The analysis process clearly identified the critical failure modes (safety and outage effects); thus the PM task decisions were focused on these failure modes. We estimate reductions in the 40 to 60 percent range, or better.
2. The reduced CM activity has naturally been of benefit in reducing unscheduled downtime. It is difficult to estimate what this might be since other facility improvements were made concurrent with the RCM process. But even small reductions here translate into meaningful cost avoidance (with downtime cost as an impact to an ongoing test estimated at $4000 per hour).
3. The tangential benefits realized from this Classical RCM process may well turn out to be a very major factor in the economic picture. While hard to estimate, the pilot project Items Of Interest (see examples below) indicate that these findings could equal or exceed the maintenance savings on some systems. In fact, these added findings are one of the "surprise" results that made us feel so very positive about the Classical RCM approach. Without such a comprehensive analysis process, it is difficult to see how this could occur to any great extent.

Overall, we intend to select systems for the RCM program which potentially could provide at least a 100-fold return on investment. Where such potential does not exist, we will use some other less costly method to review the current PM structure on the "well-behaved" systems.

One of the major benefits derived from using the Classical RCM approach was the detailed knowledge about the system design and operation gained by the team members themselves. Thus, the RCM team became experts on the system in addition to their depth of maintenance experience on individual equipment. Another fallout from the analysis is what are called Items Of Interest (IOIs) discovered during the process. In fact, these "surprise" findings yielded economic returns that far exceeded the basic cost of the RCM analysis.

These highlights alone surely give us a positive view of the Classical RCM approach. By using what some are calling a streamlined approach to RCM, which is not as comprehensive, we fail to see how a team could possibly discover IOIs of equal value. Once the Carousel Subsystem was completed, there were 38 IOIs turned over to maintenance supervisors and facility engineers for appropriate correction, documentation, evaluation, and action. A few selected IOI examples are shown in Figure 12.44.

IOI - DESCRIPTION	COMMENTS
• Paint/Coat Carousel Drive Steel Wheels to prevent Corrosion	Rapid Temp changes due to Pressurization of Six Atmospheres and Blowdown to atmospheric conditions enhance corrosion due to condensation.
• Install Dust Cover on Carousel Drive Clutch Assembly Inspection Ports	Two Large 5"Holes were cut into clutch housing for inspections. Inspection dust covers will prevent clutch contamination and excessive wear.
• Drawing M427 (design) shows Carousel location locking pin sliding sockets anchored in place using ¾" square stock and welded on three sides. As build drawings do not. They are not installed.	Square welded stock needed to guarantee no movement of pin sockets after initial installation. Stock will be welded in place and drawings corrected. A failure due to socket movement could cause downtime with an estimated productivity loss valued at $4,000 per hour.
• What procedures are in place to preclude the replacement of a failed limit switch with one that has not been properly vented?	Limit switches were found to fail as supplied unless a vent hole was drilled to equalize pressure in the switch. Procedures must be in place to eliminate wrong replacement of switch.
• It would be advantageous to eliminate the Proximity Switch from any role in a locking pin retraction scenario.	A proximity switch failure (loss of or erroneous signal) would preclude the retraction sequence; but otherwise has no role in retraction. Change Software to eliminate switch activity in circuit of the extraction mode.

Figure 12.44 Items of interest (IOIs).

Ames Research Center Management View of RCM

NASA-ARC has put a large investment into restoring the 12 ft PWT and modernizing its other high-usage wind tunnels in order to improve productivity and data quality. However, with the trend toward reducing budgets within government, the resources for wind tunnel operations and maintenance will not be available at the same levels as in the past. Therefore, we must develop new methods to optimize the utilization of maintenance resources in order to minimize both downtime associated with equipment failure and overall costs for our wind tunnel test programs. We believe that the RCM approach to PM provides that methodology to optimize our PM program.

We first learned of RCM when one of our support service contractor maintenance supervisors discovered the RCM book by Mac Smith (Ref. 1). His approach to PM, and in particular his use of what today is labeled "Classical" RCM, just made common sense to us. Thus, we embarked on the pilot project reported in this paper, and have now verified that our use of the Classical RCM method on selected critical systems, which by past standards would be considered to be maintenance intensive, is the best approach to employ.

Future Directions

- Recommendations of the RCM team will continue to be implemented in the Computerized Maintenance Management System.
- Any future corrective maintenance work on the Plenum System will be captured and evaluated against failure assumptions made by the team.
- The costs of preventive and corrective maintenance will be measured and trended as a means of evaluating the effectiveness of our maintenance program.
- The RCM team continued working other critical 12 ft PWT systems, with periodic assistance of the Consultant, and completed the RCM analysis for those systems at the end of 1996.
- Another RCM team was formed to work on critical systems at the $40 \times 80/80 \times 120$ ft Wind Tunnels, and that effort was also completed at the end of 1996.
- A third RCM team was formed in 1997, employing some members from each of the previous two teams, to work on critical systems at the Unitary Plan Wind Tunnel which was undergoing a major modernization project.

Acknowledgements

The authors wish to express their appreciation for the efforts put forth by the RCM team and supporting personnel who made this case study possible.

RCM team	*Support*
Jim Bonagofski	Herb Moss
Mike Harper	Jim McGinnis
Tony Machala	John Thiele
Dave Shiles	
Mac Smith (facilitator)	

Appendix A

LIST OF ACRONYMS

A

AE = Age Exploration

C

CB = circuit breaker
CBM = condition-based maintenance
CD = condition-directed
CM = corrective maintenance
CMMS = computerized maintenance management system
CUF = TVA Cumberland fossil plant

D

DT = downtime

E

ELM = equipment lifetime management
EPA = Environmental Protection Agency
EPRI = Electric Power Research Institute
EVA = economic value added

F

FDF = failure density function
FF = failure-finding
FGD = flue gas desulfurization

I

IOI = items of interest

J

JIT = just in time

K

KSC = Kennedy Space Center

M

MO = maintenance outage

N

NRC = Nuclear Regulatory Commission

NASA = National Aeronautics and Space Administration

O

O&M = operations and maintenance
OEE = overall equipment effectiveness
OEM = original equipment manufacturer
OSHA = Office of Safety and Health Administration

P

PDF = probability density function
PdM = predictive maintenance
PM = preventive maintenance
PMP = preventive maintenance program
PUC = Public Utilities Commission

R

RAV = replacement asset value
RCFA = root cause failure analysis
RCM = reliability-centered maintenance
ROI = return on investment
RTF = run to failure

S

SWBS = system work breakdown structure

T

TD = time-directed
TDI = time-directed intrusive
TPE = total productive engineering
TPM = total productive maintenance
TPR = total plant reliability
TQM = total quality management

W

WCM = world class maintenance
WIIF = what's in it for me

Appendix B

THE MATHEMATICS OF BASIC RELIABILITY THEORY

B.1 INTRODUCTION

In Chapter 3 we discussed some fundamental notions associated with the reliability discipline, and noted in particular the probabilistic or chance element that is basic to its understanding. We also discussed, in very simple and qualitative terms, the mathematical aspects of probability and how this is employed to derive some key elements of reliability theory which are germane to RCM.

In this appendix, we will discuss the derivation of the key elements in reliability theory in mathematical terminology. This discussion is still kept relatively simple, but a cursory knowledge on the part of the reader of some basic differential and integral calculus and probability theory will be helpful to his or her understanding of the process.

B.2 DERIVATION OF RELIABILITY FUNCTIONS

In Chapter 3, *reliability* was defined as the probability that an item will survive (perform satisfactorily) until some specified time of interest (t). We can think of reliability, then, as the expected fraction of an original population of items which survives to this time (t). Notice that the number surviving can never increase as time increases; thus, reliability must decrease with increasing time *unless* something can be done to essentially return or restore the population items to their original state which existed at $t = 0$. This, of course, is one essential aspect of preventive maintenance actions.

In deriving the reliability functions, we will consider a large number of like items on test and being run to failure. We can thus define the following parameters of interest.

Let N_0 = Original population size which will be put into operation at $t = 0$. N_0 is a *constant*, a fixed number for the population at t_0.

N_s = Population items surviving at t_x. N_s is a function of time.

N_f = Population items failed at t_x. N_f is a function of time.

$R(t)$ = Reliability of the population as a function of time.

$Q(t)$ = Unreliability of the population as a function of time.

So, at any time t_x,

$$N_0 = N_s + N_f \tag{B.1}$$

$$R(t) = N_s / N_0 = N_s / N_s + N_f \tag{B.2}$$

$$Q(t) = N_f / N_0 = N_f / N_s + N_f \tag{B.3}$$

and

$$R + Q = N_s / N_0 + N_f / N_0 = N_0 / N_0 = 1 \tag{B.4}$$

That is, R and Q are complementary events; the population items have either survived or failed at time, t_x.

$$R(t) = 1 - Q(t) \tag{B.5}$$

Since N_s must decrease with increasing time, N_f must increase.

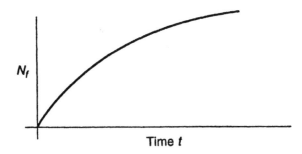

This simple plot depicts the cumulative failure history of our population over time. We can divide N_f by N_0 (recall N_0 = constant) and the basic shape of the curve does not change. Furthermore,

$$N_f / N_0 = Q(t) \tag{B.6}$$

So we now have a curve of Q versus t.

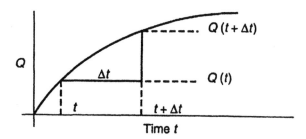

In probabilistic terminology, the curve is the cumulative density function, or CDF, of the population failure history. Also, the derivative of a CDF is a population density function, pdf—or, in this case, a failure density function, fdf.

Taking the derivative of the CDF yields the following:

$$\lim_{\Delta t \to 0} \frac{Q(t + \Delta t) - Q(t)}{\Delta t} = \frac{dQ}{dt} = \frac{d}{dt}\left(\frac{N_f}{N_0}\right)$$

$$= \frac{1}{N_0}\frac{dN_f}{dt}$$

Thus, we know that dQ/dt is a failure density function, and let $dQ/dt = f(t)$.

So,

$$f(t) = \frac{1}{N_0}\frac{dN_f}{dt} \tag{B.7}$$

From Eq. (B.7), we see that in some time interval Δt there will be some fraction of the total failures, ΔN_f, that will occur. These failures, of course, are from the original fixed population N_0, and $\Delta N_f / \Delta t$ represents the total failure frequency in Δt. When this is divided by N_0, the resulting value represents the failure frequency per item in Δt with respect to the *original* population. We call this value the *death rate*.

Thus,

$$f(t) = \text{death rate} = \frac{1}{N_0}\frac{dN_f}{dt} \tag{B.8}$$

A simple example of the death rate calculation was described in Sec. 3.4.

From here, we can perform various manipulations with the preceding equations to obtain some additional reliability functions of interest.

Taking the derivation of Eq. (B.5) yields:

$$\frac{dQ}{dt} = -\frac{dR}{dt} \tag{B.9}$$

Since $dQ/dt = f(t) = (1/N_0)(dN_f/dt)$ from Eq. (B.7), we know that

$$-\frac{dR}{dt} = \frac{1}{N_0}\frac{dN_f}{dt}$$
(B.10)

Multiplying both sides by N_0/N_s yields

$$-\frac{1}{R}\frac{dR}{dt} = \frac{1}{N_s}\frac{dN_f}{dt}$$
(B.11)

From Eq. (B.11) we see a similarity with Eq. (B.7), the death rate. But now, $\Delta N_f/\Delta t$ is divided by N_s, and the resulting value represents the failure frequency per item in Δt with respect to the population *surviving* at the beginning of the interval Δt. We call this value the *mortality* or *failure rate*, and assign the symbol $h(t)$ or λ to it. Section 3.4 also gave a simple example of the failure rate and how it is distinguished from the death rate.

Thus,

$$h(t) = \lambda = \frac{1}{N_s}\frac{dN_f}{dt}$$
(B.12)

Also,

$$h(t) = \lambda = -\frac{1}{R}\frac{dR}{dt}$$
(B.13)

Rearranging,

$$\lambda dt = -\frac{dR}{R}$$

Note that at $t = 0$, $R = 1$.

Integrating,

$$\int_0^t \lambda dt = \int_1^R -\frac{dR}{R} = -\ln R$$

And

$$R = e^{-\int_0^t \lambda dt}$$
(B.14)

Equation (B.14) is the most general formulation for reliability. No assumption has been made regarding any specific form for λ and how it varies with time.

Recapping, if we know the failure density function $f(t)$, we can derive all other reliability functions of interest. We thus see the importance that can be attached either to our ability to experimentally determine $f(t)$, or to credibly assume some form of $f(t)$.

B.3 A SPECIAL CASE OF INTEREST

The failure density function $f(t)$ most often used in reliability analyses is the exponential fdf, which takes the form:

$$f(t) = \lambda e^{-\lambda t} \tag{B.15}$$

In Eq. (B.15), λ is a *constant* value, and thus, for any Δt of interest, λ is a constant. That is to say that the mortality or failure rate is a constant, so the λ in Eq. (B.15) is also our λ which we derived as the failure rate.

Or, if you wish, we could assume that λ in Eq. (B.14) is a constant and work backwards to obtain $f(t)$:

$$R = e^{-\int_0^t \lambda dt} = e^{-\lambda t}$$

$$-\frac{dR}{dt} = -\frac{d}{dt}(e^{-\lambda t}) = \lambda e^{-\lambda t}$$

But,

$$-\frac{dR}{dt} = \frac{dQ}{dt} = f(t)$$

So, when λ is a constant, the corresponding $f(t)$ is

$$f(t) = \lambda e^{-\lambda t}$$

When λ = constant is assumed (or known), the implications, in hardware terms, are important to understand:

1. The failures in any given interval of time, on average, occur at a constant rate. These failures are random in nature—that is, we really don't know just what failure mechanisms are involved or what causes them and, consequently, we do *not* know how to prevent them!
2. If we believe (or know) that λ = constant for the items in question, but we do not know the specific value of λ, we could test 1000 items for 1 hour or a few samples for 1000 hours, and calculate λ. Either way, the resulting values would be approximately the same (if our λ = constant assumption is truly correct).
3. The mean of the exponential fdf, or the mean time to failure (MTTF), is $1/\lambda$. Thus, when the elapsed time of operation is equal to the MTTF:

$$R = e^{-\lambda t} = e^{-\lambda(1/\lambda)} = e^{-1} = 36.8\%$$

4. From 3, we can further understand that when the accumulated operating time is equal to the MTTF, there is a 63.2 percent chance that a randomly selected item in the population has already failed.

Notice that items 3 and 4, when clearly understood, tell us that it is really not wise to use the MTBF as a guide for determining PM task frequency. Among the other pitfalls associated with the λ = constant case, this is one of the unfortunate myths that maintenance organizations often employ.

There is one additional aspect that we should discuss. Every fdf, no matter what form it might take, will have a mean value or, in reliability terms, a mean time to failure (MTTF). In the exponential case, there is an MTTF which we can further label as MTBF due to the λ = constant property.

But suppose $\lambda \neq$ constant.

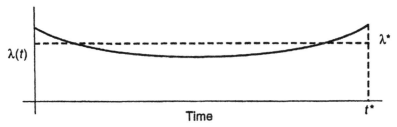

In theory, we can find an average value of λ^* where the area $\lambda^* \cdot t^* = \int_0^{t^*} \lambda(t)dt$. Thus, we could consider λ^* to be a "constant" from $t = 0$ to $t = t^*$, and call $1/\lambda^*$ an average MTBF. In practice, this is frequently done—but without really knowing if we are dealing with a true λ = constant case.

This could be dangerous. Consider, for example, a slightly different picture from the one just shown.

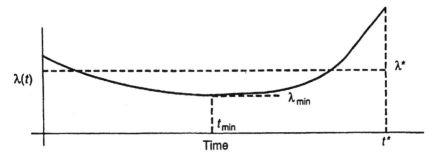

If the equipment operates to t_{min}, then our estimate of λ^* is conservative, and we would experience failure rates *less* than expected. But if we operate to t^*, we would experience failure rates considerably larger than expected (perhaps by 2× or 3×), and this could be very devastating!

Again, the importance of knowing $f(t)$, and whether λ varies with time or not, becomes evident when dealing with equipment and system reliability issues.

Appendix C

THE ECONOMIC VALUE OF PREVENTIVE MAINTENANCE

David Worledge, Applied Resource Management

OVERVIEW

A PM optimization project usually has its origin in the recognition of deficiencies in the maintenance processes, or in the performance of one or more plant systems. To focus the effort where it will bring the optimum return on investment requires scoping activities which have been summarized in Chapter 1, Sec. 1.4.4 and discussed in detail in Chapter 5. These activities usually make use of parameters such as direct maintenance costs, the system contribution to forced outage rate, and plant downtime. When data is lacking to support such investigations, rough estimates or subjective judgements of these and other parameters can often be made by experienced individuals—although they may be biased by internal organizational conflicts and incomplete knowledge of certain aspects of operations and maintenance.

For example, a machine may experience a lot of failures and downtime, requiring a mechanic and an electrician to be assigned practically full time to keep it productive. Although it is a poor performer, and may have a clearly deficient PM program, this asset may not improve impressively when an RCM-based program is implemented. The reason could be that many of the failures are caused by software errors, or operator error, or by intrinsic machine design inadequacies. Alternatively, much of the downtime may be due to logistical problems related to the supply of parts, tools, materials, or the availability of other equipment such as cranes.

Downtime may not be a serious problem if the machine or process can be reset and restarted relatively quickly, so these stoppages may be perceived as relatively benign. However, if interruptions to the manufacturing process lead to a need to rework or even to scrap a percentage of the output, a lack of understanding of the true cost of the rework will disguise the importance of reliability. In one case experienced by the author, management judged the rework cost to be the cost of resubmitting the defective parts to the same machine—which turned out to be a relatively minor cost. Further investigation revealed that the administrative costs of tracking the reworked parts and regenerating the QA paperwork was seven times greater than the apparent operational costs. Further investigation also showed that the rework caused regular production to be pushed into overtime hours, and that a lack of buffer storage between the machine and the next process on the production line required the next stage of the process to be staffed at over-time rates as well. In all, the initial reliability problems, which caused little down-time, nevertheless had an outsized effect on production costs, requiring twelve times more workers to be paid overtime than the number required to simply run the subject machine, all in addition to the QA costs. Unfortunately, it is not likely that sufficient capability and resources will be devoted to the screening step to uncover these all too common anomalies early in the project.

Once an RCM project has been completed, management typically expects an esti-mate of the likely benefit—assuming that the recommended PM changes are fully implemented. Of course, RCM practitioners are well known for their claims that optimizing PM can bring benefits of the order of 30 to 70 percent reduction in direct maintenance costs and downtime. These numbers are certainly based on actual case history experience, but often fail to convince plant management because they are not plant-specific estimates. Indeed, the benefits could be close to zero, or close to 100 percent, depending on the condition of the original PM program and the kind of factors identified above. In a typical case the direct PM costs may increase substantially, but a 30 percent improvement in downtime is almost always more than enough to quickly dominate the results. However, even in such a case of rapid cost recovery, it may not be obvious that the RCM project was the best use of resources. For example, if a significant fraction of the origi-nal failures was not maintenance preventable, for whatever reason, the resources spent on RCM might have been more profitably devoted to the other issues, even though the RCM project can more than repay its own costs.

At this point one thing should be clear. Capturing the main influences to deter-mine the best use of resources, even approximately, requires us to push towards further understanding of the costs of failures and downtime. This can be accom-plished using software which would embody a reliability-maintenance model and a production-cost model. The tactical objective would be to assign a credible monetary value to PM activities and to the PM program, but, if successful, there could be additional benefits, including a critical strategic payoff (ROI). Such benefits would be the ability to deploy maintenance resources more rationally, and to effect and measure program improvements, in terms of company profits on a per-asset basis. The strategic benefit would be to shift management perceptions of maintenance from being mainly a cost that is a drain on company resources, to the more positive ground of a value-added activity. In other words, treat the maintenance organization as a *profit center* as proposed in Sec. 1.5. The pay-off will, therefore, not be limited to increased production at lower cost. Credible portrayal of the effects of maintenance in dollar and profit-oriented terms will lead management increasingly to recognize the leverage provided by the PM program, and the value to the company of those who implement it.

This strategic benefit is potentially of enormous importance to the maintenance organization. Preventive maintenance is not a topic with much appeal to senior management of most companies. Maintenance of plant and facilities typically consumes sufficient resources to be a significant cost of operations. Yet the positive return on this investment is not readily discernible. Corrective maintenance to repair equipment after forced outages certainly can be seen to have an important restorative effect, but management will see this, correctly, as a negative (i.e., a cost) feature. On the other hand, preventive maintenance often has no immediate and visible benefit at all. (It is difficult to take credit for something that did not occur.)

Even though most managers who lack a maintenance background are aware of the need to perform some PM tasks to help keep production at the desired level, PM program improvements have to compete with other priorities, many of which will have conventional cost justifications. The problem is that a quantitative link between PM resource expenditure and improved production is usually not demonstrable before the fact, or even over a short period of time after the fact. Worst of all, the perception is in the form of a double negative: resources spent on PM negate the negative effects of equipment breakdowns—not exactly a ringing endorsement which will ensure success in competing for company resources.

THE REPRESENTATION OF VALUE

Value is usually revealed by cost–benefit analysis. In the case of PM, it is important to capture the costs of equipment failures, to distinguish critical failures which cause asset downtime as well as safety or quality problems, and to follow the cost effects of these events far enough to represent most of their true impact.

Clearly a *production-cost model* is needed which attempts to represent the interaction between production and downtime. Such interaction must incorporate production targets during regular shifts, and the ability to make up for lost production due to all causes of equipment downtime, including the need to reprocess some of the product and to replace product that has to be scrapped.

The way that lost production is replaced in overtime shifts, and the potential for manufacturing new product during overtime, must all be included.

A production-cost model will capture the revenues due to uptime, and the costs due to downtime, including the direct cost of maintenance. This kind of model, with actual production and cost data for given periods, can be used to deduce the economic value of each asset and to track its performance. Such a model is relatively conventional, although it can easily outstrip the needs and capability of a maintenance organization in terms of complexity.

Although a production-cost model can provide insights into the different production and maintenance cost drivers, there is still nothing in such an approach to show the *benefit* of PM. To be beneficial, a model of PM value must also project the effect of PM activities on reliability and downtime. Then a change in PM will generate a change in the number of failures, a change in downtime, and a change in the amount of rework. When coupled with the production-cost model, the change in production costs and income can be estimated. These are the data inputs required for a cost–benefit evaluation of the PM change.

Thus, representing the impact of PM on reliability and downtime requires a *reliability–maintenance model*. This model must take account of the proportion of all failures which are maintenance preventable (e.g., those failures caused by software faults or operator error are not affected by PM improvements), the proportion of these which are critical (e.g., all failures must be repaired, but only the critical ones may cause downtime or rework), and the proportion of critical and non-critical failures which are protected by PM task(s) to certain levels of effectiveness. Additionally, the model should account for the lost opportunity costs of the lost production. In cases of high demand (e.g., times when overtime production is required to meet current customer demand), these costs can be a significant, or even dominant, cost associated with lost production.

Remarkably, it has been found that most of this information can be generated with little effort during the course of a regular RCM evaluation, or is directly derivable from the results. Even better, given the structured input of operators, maintenance personnel, and data from a TPM program (such as Overall Equipment Effectiveness data if it exists), it is not difficult to estimate values of the data inputs before maintenance improvement is attempted. These approximate inputs can be quite sufficient to add considerable intelligence to the screening of assets to select those where PM optimization will be the best use of resources. Refinement of the data after the RCM analysis will improve the results, and will establish a basis for longer term tracking and trending of asset performance and maintenance effectiveness.

In the preceding discussion, the term "model" has been used to emphasize that this is not an accounting procedure which uses precise data to produce precise results, but an engineering evaluation which makes relevant approximations. This approach reduces the required data, and keeps focus on what we really need to know: (1) would the PM change produce a significant improvement in reliability and downtime; (2) what value would this add to the company in relation to the current value added by the asset; (3) what other areas might be addressed before PM is improved; and (4) what should be measured to monitor asset value and maintenance effectiveness going forward?

It is easy to become discouraged by all the uncertainties in such an endeavor, but if we make no effort to quantify our experience of the relevant issues, our resource allocation decisions will be in danger of not taking these influences into account at all. The intent of all the techniques discussed in this book (e.g., RCM, Living PM Programs, production-cost models) is to allow management to make informed decisions concerning facility maintenance programs, production processes, and resource allocation. It should be remembered that these decisions will be made regardless of the use of these tools. The tools proposed here represent structured methods that provide management personnel with all the relevant and available information, in a consistent manner. Using these tools, decision makers will make better quality resource decisions with a higher level of confidence.

GENERAL SCOPE OF THE MODELS

To achieve the objectives discussed previously, we must generate a compact view of the economic value added by any company asset, and its sensitivity to changes in PM. The models must provide a summary of production and maintenance costs, throughput, and revenue, and give a bottom-line report on the asset's economic value to the company. Statistical data on quantities such as downtime, throughput, and the number of preventive and corrective maintenance man-hours can be used to represent prior performance over an arbitrary period of time—which can be used as a monitoring period.

First, we need to calculate results for the current PM program, i.e., that for which the input statistics apply. Cost categories would include PM labor cost, PM spare parts cost, breakdown labor and spares costs, and the costs to make up for production lost in various categories of asset downtime, depending on whether the makeup production occurred at regular or overtime shift rates. We would include normal operating and materials costs, and calculate revenue and pre-tax operating profit for the particular asset. The after-tax operating profit minus the cost of capital could be used to represent the value added by the asset. This quantity has been referred to as the Economic Value Added (EVA).

The EVA has to be calculated for specific assumptions regarding production. For example, one option would be to assume that production during regular (weekday) shifts is the primary production mode, with overtime shifts used primarily to make up for production lost during the week. Excess overtime hours, if available after making up for lost production, may or may not be used to manufacture additional parts or product.

The second part of the calculation would exercise the reliability–maintenance model to estimate the impact of a change in the PM program. It would recalculate all the above quantities for the proposed (i.e., changed) program. Tables of results would compare the new program with the old in absolute and percentage terms, and chart key items. This would be ideal for investigating the prospective, and actual, benefits of performing an RCM optimization.

Software with exactly this scope and capability has been prepared to investigate the feasibility, credibility, and practical value of the approach. The software is called ProCost© and has been implemented as a Microsoft Excel™ 97 program which produces a workbook for each calculation. This capability is suitable for evaluation of PM program changes on a small number of assets. The software has also been written in a Microsoft Access™ database version which archives the input data and the results of periodic calculations. The results can be configured to trend and review the performance of a large number of assets in convenient subsets, e.g., by asset type, by process facility, or by company sector, e.g., by region. Versions are available which address the needs of manufacturing facilities, as described in this appendix, and also electricity generating plants.

DATA DEFINITION ISSUES

The data needs of the models were constrained to be a good fit to data systems available at the plant where the first implementation of ProCost© was made. This implementation was done at a large Midwestern manufacturing and assembly plant. In the late 1990s, this company embarked on a restructuring of the maintenance organization and a redesign of its maintenance processes and procedures as part of a wider effort to enhance production and reduce costs. The maintenance organization intends to use the ProCost© tool to focus its future RCM evaluations on the assets showing the largest value-added from such improvements, and to ensure that all assets continue to add value to the bottom line.

Even after the model parameters were redefined to better fit the plant's data capabilities, issues of data reporting and definition remained. It is common experience that data quality depends on the understanding and compliance of the employees reporting the individual data values. However, the challenges of data definition go beyond the issues of training and historical usage in the facility. The definitions for the terms "preventive" and "corrective" maintenance require careful consideration in order to be stated with precision and clarity, and the operational implementation of these definitions could vary somewhat between industries. It also is vitally important to decide which work orders are of each kind. To this end,

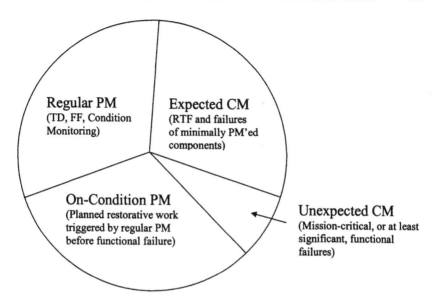

Figure C.1 Partition of work orders.

Figure C.1 shows a partition of all *work orders* which address both healthy and degraded or failed components. This figure does not characterize the types of PM tasks; see Chapter 2 for a complete discussion on PM task definition. Instead, we are looking at the work that is done under various types of work orders at this particular plant, and dispositioning the nature of their work as Preventive Maintenance or Corrective Maintenance.

Work orders that address the regularly scheduled PM tasks are labeled "Regular PM." These are the work orders that implement traditional time-directed PM tasks such as inspections, and restore/replace activities, failure-finding tasks such as surveillance tests, and condition-monitoring, performance-monitoring, and other predictive maintenance activities.

The category "On-Condition PM" refers to work orders in which degraded sub-components, discovered during the execution of regular preventive maintenance tasks on the main component, are repaired or replaced. If this restorative work is carried out at a later date, it is typically performed under what most facilities regard as "corrective maintenance" work orders. But these degraded subcomponents (some may even be failed) *are fully anticipated by the PM program*, so the sub-component degraded conditions or failures do not per se immediately constitute the larger impact or loss-of-important-function failures which the PM program is designed to prevent. An example would be tightening the packing on a pump after a leak is discovered during a routine inspection, provided the leak does not limit the function of the pump. A second example would be the planned changing of a motor bearing after high vibration is discovered during vibration monitoring. In this case, the motor could have a very important function, but the emergent condition is corrected by *planned* intervention before failure occurs. In total, there is a large number of these activities where the work of correcting degraded conditions (which were implicitly anticipated) is not performed precisely during the On-Condition discovery PM task.

The insistence that the triggered and emergent work be planned before being considered to be a part of PM places significant constraints on the effectiveness of condition-monitoring tasks. If the emergent work is so urgent that it forces a high impact outage, it obviously has to be interpreted as true corrective maintenance. The word "planned" implies there is adequate time to properly plan the work so that the outage can be taken at a time when it minimizes loss of function. This planning may often require the use of Age Exploration to ascertain just how this can be accomplished (see Sec. 5.9).

The "Expected CM" work includes the run-to-failure cases, which at this Midwestern plant require corrective maintenance work orders to repair them. But these failures are expected to occur, and they are an anticipated aspect of the PM program. It is not clear that these work orders should be classified as corrective maintenance work orders because they form a class of expected corrective maintenance that does not indicate a poor PM program, a class which could indeed be increased rather than decreased by RCM optimization. In a similar way, we should also include among the "Expected CM" work that which is required to repair failures of the components that receive only minimal PM. To the extent that some PM is indeed performed on this equipment, some of these failures are, in fact, unexpected, but the majority will be associated with failure modes that are not by choice protected by PM. It will not be cost effective to separate the two types of work orders for this category of equipment whose failures have minimal impact. Classing all of these failures as "Expected" also emphasizes that they have been planned and anticipated by the PM program.

Finally, there are the true functional failures which constitute the more costly events that PM tries to prevent. These can claim to be "Unexpected," and their repair can be labeled as "Unexpected CM."

In any application where the PM and CM distinction is relevant, such as the estimation of the costs of unreliability, it is important to classify work orders properly so that those addressing the On-Condition work are included with the regular PM events on the PM side of the costs. Only part of this requirement can be met by careful process design. Training is also required, as inadequate personnel training on data reporting will result in incorrect classifications, thus limiting the utility of the model. For example, a common problem is the reporting of true corrective work on a preventive work order because the opportunity is taken to perform a restoration task on an unacceptable as-found condition. It also seems to be true that even if someone is assigned to review all work orders, some PM/CM categorization decisions require considerable experience, usually because of uncertainty over the level of functional impairment, or the degree to which On-Condition work was really planned and avoided a forced outage. The costs of expected CM at this facility were treated as part of corrective maintenance in ProCost©. Even in a perfect PM program which eliminates all unexpected CM, there will therefore remain a significant CM cost, consisting of the expected contributions from running to failure the functionally unimportant components, and repairing those failures with minor economic impact.

The result is that we should anticipate that there will always be some CM cost, even in a perfect PM program, and even when the On-Condition costs are properly allocated to the PM program. The issue of whether to treat the expected CM costs as CM or PM is illuminated by this discussion. Treating them as CM, as ProCost© does, acknowledges the fact that they are repairs of failures, albeit anticipated and relatively inconsequent ones. Adding their cost to the other CM costs does not distort the effectiveness of the PM program, because the PM program should be designed to minimize the total cost by providing an appropriate balance between preventing failures and allowing them to occur. It is an important distinction to make: the PM program should minimize the total maintenance cost, not just the corrective maintenance cost. (See the discussion in Sec. 10.3 which expresses the same conclusion.)

METRICS

ProCost© calculates eleven quantities to track aspects of performance meaningful to a maintenance organization. These metrics are calculated using regular shift work as a standard basis for value added by the asset, and for maintenance effectiveness. Including the overtime shifts can distort the data because, on some assets, the overtime operating crews may not be as familiar with the equipment as regular crews, and other logistical problems may occur which are not typical of normal operating conditions. This is an example of where we need to keep the focus on showing the effectiveness of maintenance, rather than calculating a complete picture for the accountants. The metrics focus on the areas of unavailability, the amount of PM and CM, throughput and rework, the cost of production losses, and the economic value added to the company by the asset.

Three measures of unavailability explore the fractional downtime which is caused by different activities and organizations. *Maintenance unavailability* has contributions only from asset outages caused by doing PM and by the completion of repairs that are maintenance preventable, i.e., true corrective maintenance. The maintenance organization "owns" this unavailability. *Machine breakdown unavailability* has contributions caused only by machine breakdowns, but these can be due to both hardware and software faults, the latter not being the responsibility of the maintenance organization. *Operations unavailability*, such as waiting on parts or tools, is also not the responsibility of the maintenance organization, but is often an even larger quantity than the first two metrics. These three parameters easily could be redefined to suit somewhat different circumstances, but each tells a tale, and carries a message for a certain group of individuals.

The next two metrics are man-hour parameters which essentially track the amounts of PM and CM. Then there are three metrics which attempt to bring to everyone's attention the "true" costs of failures and downtime. The first of these is the *Cost to make up lost production* which displays the dollar cost attributable to all failures and downtime, where:

Makeup cost ratio =

$$\frac{(\text{Costs to make up all losses} + \text{Regular shift operator labor} + \text{Materials})}{(\text{Regular shift operator labor} + \text{Materials})}$$

This represents "value thrown away." The other two metrics are slightly different ways to compare the actual cost to make the parts with what they would have cost if there had been no failures or downtime. Finally, the *Economic Value Added* provides the bottom line as to how much money the asset is making or losing for the company, given by:

Economic Value Added = After-tax Operating Profit − Cost Of Capital

In summary, the ProCost© software enables the user to estimate production and maintenance costs, throughput, revenue, economic value added, and other metrics for a specific asset, using engineering models that contain suitable engineering approximations. The main idea is to create standardized measures of asset performance using the models and statistical data, with a focus on the value added by preventive maintenance. ProCost© is designed to serve the needs of reliability engineers, maintenance planners, maintenance engineers, and facilities management. Although the results hold considerable interest for production or accounting personnel, the current version is not designed to specifically serve their needs.

TYPICAL RESULTS

The calculation demonstrated here is for one of the large drilling and routing tools in the company's Midwestern facility. These machines were manned by two operators and were run continuously on a three-shift basis. They were experiencing continuous breakdowns, to the extent that a maintenance mechanic and an electrician were spending essentially all of their time on just two machines.

This Section Data Inputs Are Complete								
Planned administrative down time	Total	697	Reg.	496	OT			(hours)
Unplanned administrative down time	Total	441	Reg.	314	OT			(hours)
Software breakdown downtime	Total	165	Reg.	117	OT			(hours)
Hardware breakdown down time	Total	662	Reg.	471	OT			(hours)
Setup time	Total	634	Reg.	451	OT			(hours)
Breakdown labor man hours	Total	2284	Reg.	1626	OT			
PM labor man hours	Total		Reg.	24	OT	9		
Breakdown spares cost	Total	$137,997	Reg.		OT			
PM spares cost	Total	$1,310	Reg.		OT			
Scheduled downtime for doing PM	Total		Reg.	15	OT	6		(hours)

Reg. means regular shifts, OT means overtime shifts. For each parameter, any two of the three inputs is required, since Total = Reg. + OT.

Figure C.2 Statistical data inputs for the sample asset.

Over time, the existing PM program had deteriorated, probably because it was not well designed initially, and the large amount of breakdown maintenance had pre-empted preventive activities.

This was therefore a relatively simple case where there was little question that a better PM program had to be developed. Most of the hardware failures experienced were judged to be maintenance preventable, which encouraged this view. However, the machine was subject to a moderate amount of operator error and software faults, and there was a significant amount of administrative downtime. These remaining ills diluted the benefits from PM optimization. The RCM analysis revealed that about 40 percent of the critical failures had not previously been protected by any kind of preventive tasks. Figure C.2 shows the statistical data input.

The bar chart which follows, Figure C.3, shows a comparison of the major results projected by ProCost©. The PM program change is shown to be very effective in reducing direct maintenance costs and significantly reducing total production costs. In turn, this increases income and pre-tax operating profit, and permits a positive value to be generated by the asset. Observe the reversal of the small excess of Total Production Costs Over Total Income. Notice also the effect of taxes and the cost of capital which together significantly reduce the improvement in Pre-Tax Operating Profit and result in the smaller improvement visible for Economic Value Added. Even with these reductions, the EVA becomes a gain of $115,000 per year instead of a loss of $315,000 per year—a marked contrast to the prior (i.e., the existing) situation.

Figure C.4 provides additional breakdown of the results. The RCM project was quite successful in reducing direct maintenance costs at a small level of PM expenditure. The value to be obtained from spending one additional dollar on

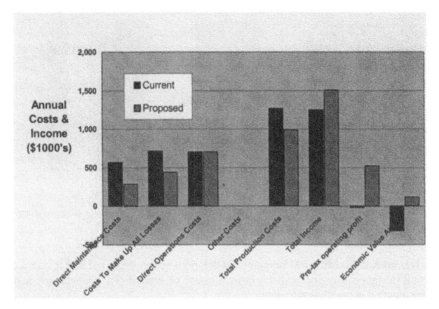

Figure C.3 Comparison of major cost and value categories for the current and proposed PM programs.

Category	Current	Change
PM labor	3,497	15,171
PM spares	3,001	15,915
Breakdown labor	244,339	-135,300
Breakdown spares	316,139	-175,058
Total Maintenance Direct Cost	566,976	-282,274
Total EVA	-315,118	430,946
PM leverage		x 15.34

Figure C.4 Direct maintenance costs, EVA, and leverage of the PM program.

improved PM is calculated to be just over $15. This is a large number; it demonstrates that properly applied PM is indeed a money maker, and can directly improve the bottom line for a very modest expenditure of company resources. Notice that the annual increase in PM costs is about $31,000, under 6 percent of the current direct maintenance cost of breakdowns.

However, this asset still experiences large losses from the combination of software and logistics problems which prevent it from reaching its profit potential. Figure C.5 shows the projected metrics for the proposed (optimized) PM program

METRICS FOR DATA INPUT PERIOD			
	Current	Proposed	Proposed minus Current
Maintenance Unavailability (R)	22.8%	7.6%	-15.2%
Machine Breakdown Unavailability (R)	27.6%	12.3%	-15.3%
Operations Unavailability (R)	14.7%	18.0%	3.2%
Hardware Breakdown Man-Hours (T)	2284	1019	-1265
Preventive Maintenance Man-Hours (T)	33	175	142
Number Of Good Units Produced (R)	1502	1796	295
Units Needing To Be Reprocessed (R)	0.0	0.0	0.0
Number Of Units Scrapped (R)	6.4	4.5	-1.9
Operations Makeup Cost (R)	$713,531	$439,334	-$274,197
Makeup Cost Index	3.53	2.32	-1.21
Average Cost Ratio	3,82	2.29	-1.54
Economic Value Added (T)	-$137,552	$50,560	$188,112

In this table (R) refers to results for regular shifts, (T) to total shifts.

Figure C.5 Metrics for the sample asset.

for the data input period. The make-up costs are still much larger than the EVA, because Operations Unavailability greatly exceeds the now improved Maintenance Unavailability. The Makeup Cost Index and the Average Cost Ratio have declined significantly but still are well above the practical minimum value of around 1.5. If these calculations had been available before the RCM project, they would have added useful context to the resource allocation decisions, and might have changed the project priorities or the schedule.

CONCLUSION

The production-cost and maintenance models implemented in ProCost© give a clear view of whether an asset is producing value and, in either case, the benefit that can be gained by improving the PM program. Without such a tool, factors such as the proportion of software errors and operator errors, unplanned logistical downtime, the change in effectiveness of the PM tasks, the enhanced costs of making up for lost production during overtime shifts, the effects of taxes, and the cost of capital, can obscure the merits of PM improvement.

This easily can diminish the prospects of competing successfully for company resources. It should be clear that the ProCost© analysis provides a unique process for ranking potential company gains from improvements of different kinds to various assets.

Facility Maintenance is intending to use ProCost© in deciding which assets would benefit most from PM improvement. In addition, regular trending of relevant metrics will help keep PM for selected assets on the right track. Beyond that, the high values of PM leverage show the proactive value of the contribution made by their Facilities Services organization. Over time, this should increase awareness among all levels of management of the value-added aspect of preventive maintenance, and should help the organization to compete more successfully for company resources.

REFERENCES

1. Smith, Anthony M., *Reliability-Centered Maintenance*, McGraw Hill, 1993, ISBN 0-07-059046-X.
2. Hudiberg, John J., *Winning with Quality: The FPL Story*, Quality Resources—A Division of the Krause Organization Ltd, 1991, ISBN 0-527-91646-3.
3. Hartmann, Ed, "Prescription for Total TPM Success," *Maintenance Technology*, April 2000.
4. Ellis, Herman, *Principles of the Transformation of the Maintenance Function to World Class Standards of Performance*, TWI Press, 1999.
5. Mitchell, John S., "Producer Value — A Proposed Economic Model for Optimizing (Asset) Management and Utilization," *MARCON 98*, 1998.
6. Westbrook, Dennis, Ladner, Robert, and Smith, Anthony M., "RCM Comes Home to Boeing," *Maintenance Technology*, January 2000.
7. Koch, Richard, *The 80/20 Principle — The Secret of Achieving More with Less*, Currency Doubleday, 1998.
8. Mobley, R. Keith, *Introduction to Predictive Maintenance*, 2nd Edition, Butterworth–Heinemann, October 2002, ISBN 0-7506753-1-4.
9. Nicholas, J., and Young, R. Keith, *Predictive Maintenance Management*, 1st Edition, Maintenance Quality Systems LLC, January 2003, ISBN 0-9719801-3-6.
10. Corio, Marie R., and Costantini, Lynn P., *Frequency and Severity of Forced Outages Immediately Following Planned or Maintenance Outages*, Generating Availability Trends Summary Report, North American Electric Reliability Council, May 1989.
11. Flores, Carlos, Heuser, Robert E., Sales, Johnny R., and Smith, Anthony M. (Mac), "Lessons Learned from Evaluating Launch-site Processing Problems of Space Shuttle Payloads," *Proceedings of the Annual Reliability & Maintainability Symposium*, January 1992.
12. *RADC Reliability Engineer's Toolkit*, Systems Reliability and Engineering Division, Rome Air Development Center, Grifiss AFB, NY 13441, July 1988.
13. *Reliability, Maintainability and Supportability Guidebook*, Society of Automotive Engineers, 2nd Edition, June 1992, Library of Congress Catalog Card No. 92-60526, ISBN 1-56091-244-8.

14. Kuehn, Ralph E., "Four Decades of Reliability Experience," *Proceedings of the Annual Reliability & Maintainability Symposium*, January 1991, Library of Congress Catalog Card No. 78-132873, ISBN 0-87942-661-6.

15. Knight, C. Raymond, "Four Decades of Reliability Progress," *Proceedings of the Annual Reliability & Maintainability Symposium*, January 1991, Library of Congress Catalog Card No. 78-132873, ISBN 0-87942-661-6.

16. Nowlan, F. Stanley and Heap, Howard F., *Reliability-Centered Maintenance*," National Technical Information Service, Report No. AD/A066-579, December 29, 1978.

17. *Reliability Centered Maintenance Guide for Facilities and Collateral Equipment*, National Aeronautics and Space Administration, February 2000.

18. Matteson, Thomas D., "The Origins of Reliability-Centered Maintenance," *Proceedings of the 6th International Maintenance Conference*, Institute of Industrial Engineers, October 1989.

19. Personal communications between A. M. Smith and T. D. Matteson in the period 1982–1985.

20. Bradbury, Scott J., "MSG-3 Revision 1 as Viewed by the Manufacturer (A Cooperative Effort)," *Proceedings of the 6th International Maintenance Conference*, Institute of Industrial Engineers, October 1989.

21. Glenister, R. T., "Maintaining Safety and Reliability in an Efficient Manner," *Proceedings of the 6th International Maintenance Conference*, Institute of Industrial Engineers, October 1989.

22. *Reliability-Centered Maintenance for Aircraft Engines and Equipment*, MIL-STD 1843 (USAF), 8 February 1985.

23. *Reliability-Centered Maintenance Handbook*, Department of the Navy, Naval Sea Systems Command, S 9081-AB-GIB-010/MAINT, January 1983 (revised).

24. *Application of Reliability-Centered Maintenance to Component Cooling Water System at Turkey Point Units 3 and 4*, Electric Power Research Institute, EPRI Report NP-4271, October 1985.

25. *Use of Reliability-Centered Maintenance for the McGuire Nuclear Station Feed-water System*, Electric Power Research Institute, EPRI Report NP-4795, September 1986.

26. *Application of Reliability-Centered Maintenance to San Onofre Units 2 and 3 Auxiliary Feed-water Systems*, Electric Power Research Institute, EPRI Report NP-5430, October 1987.

27. Fox, Barry H., Snyder, Melvin G., Smith, Anthony M. (Mac), and Marshall, Robert M., "Experience with the Use of RCM at Three Mile Island," *Proceedings of the 17th Inter-RAM Conference for the Electric Power Industry*, June 1990.

28. Gaertner, John P., "Reliability-Centered Maintenance Applied in the U.S. Commercial Nuclear Power Industry," *Proceedings of the 6th International Maintenance Conference*, Institute of Industrial Engineers, October 1989.

29. Paglia, Alfred M., Barnard, Donald D., and Sonnett, David E., "A Case Study of the RCM Project at V.C. Summer Nuclear Generating Station," *Proceedings of the Inter-RAMQ Conference for the Electric Power Industry*, August 1992.

30. Crellin, G. L., Labott, R. B. and Smith, A. M., "Further Power Plant Application and Experience with Reliability-Centered Maintenance," *Proceedings of the 14th Inter-RAM Conference for the Electric Power Industry*, May 1987.

31. Smith, A. M. (Mac), and Worthy, R. D., "RCM Application to the Air Cooled Condenser System in a Combined Cycle Power Plant," *Proceedings of the Inter-RAMQ Conference for the Electric Power Industry*, August 1992.

32. *Commercial Aviation Experience of Value to the Nuclear Industry*, Electric Power Research Institute, EPRI Report NP-3364, January 1984.
33. Moubray, John, *Reliability-Centered Maintenance; RCM II*, Second Edition, Industrial Press, 1997, ISBN 0-8311-3078-4.
34. *RCM Cost–Benefit Evaluation*, Electric Power Research Institute, Interim EPRI Report, January 1992.
35. *Comprehensive Low-Cost Reliability Centered Maintenance,*" Electric Power Research Institute, EPRI TR-105365, September 1995.
36. *Innovators with EPRI Technology*, Electric Power Research Institute, Bulletin IN-105194, June 1955.
37. Moubray, John, "Is Streamlined RCM Worth the Risk?" *Maintenance Technology*, January 2001.
38. Hefner, Rod, and Smith, Anthony M. (Mac), "The Application of RCM to Optimizing a Coal Pulverizer Preventative Maintenance Program," *Society of Maintenance and Reliability Professionals 10th Annual Conference Proceedings*, Nashville, TN, October 2002.
39. Fox, B. H., Snyder, M. G. (Pete), and Smith, A. M. (Mac), "Reliability-centered maintenance improves operations at TMI nuclear plant," *Power Engineering*, November 1994.

Index

ABOUT THE AUTHORS

Anthony M. (Mac) Smith
AMS Associates

Anthony M. (Mac) Smith is internationally recognized for his pioneering efforts in the application of Reliability-Centered Maintenance (RCM) to complex systems and facilities in the industrial and government areas. His engineering career spans 50 years of technical and management experience including 24 years with General Electric. For the past 23 years, he has concentrated on providing RCM consulting and education services to many of the Fortune 100 companies, and also to the Air Force, Navy, and NASA. Mac has published more than 50 technical papers, and authored his first book on RCM in 1993 (see Ref.1). His work spans projects in the energy, aerospace, and high volume manufacturing sectors. He is an Associate Fellow of the American Institute of Aeronautics and Astronautics. Mac resides in San Jose, California, and is a registered Professional Engineer in California.

Glenn R. Hinchcliffe, PE
G&S Associates

Glenn R. Hinchcliffe is a consultant to a diverse array of clients in the energy, aerospace, government, and industrial sectors. He has over 20 years of direct experience in organizational and maintenance optimization, specializing in the application of Reliability-Centered Maintenance (RCM) and systems analysis. His background and direct experience in reliability for maintainable systems, along with his contribution to the Electric Power Research Institute's Preventive Maintenance Database, places him in the unique position of understanding the forces affecting today's maintenance professional–how the seemingly divergent goals of increasing plant availability at the least cost may be achieved. He resides in Charlotte, North Carolina where he formed G&S Associates in 1997, is a registered engineer in the states of Florida and North Carolina, and a senior member of IEEE. Mr. Hinchcliffe's life work and achievements have been recognized by the National registry of 'Who's Who'.

Printed and bound by CPI Group (UK) Ltd, Croydon, CR0 4YY

03/10/2024

01040433-0006